MY FATHERS' HOUSES

ALSO BY STEVEN V. ROBERTS

From This Day Forward

WITH COKIE ROBERTS

Eureka

Dad with the "twinnies." I'm on the left, Marc is on the right. Dad used our middle names, Jeffrey Victor, as his byline in writing children's books.

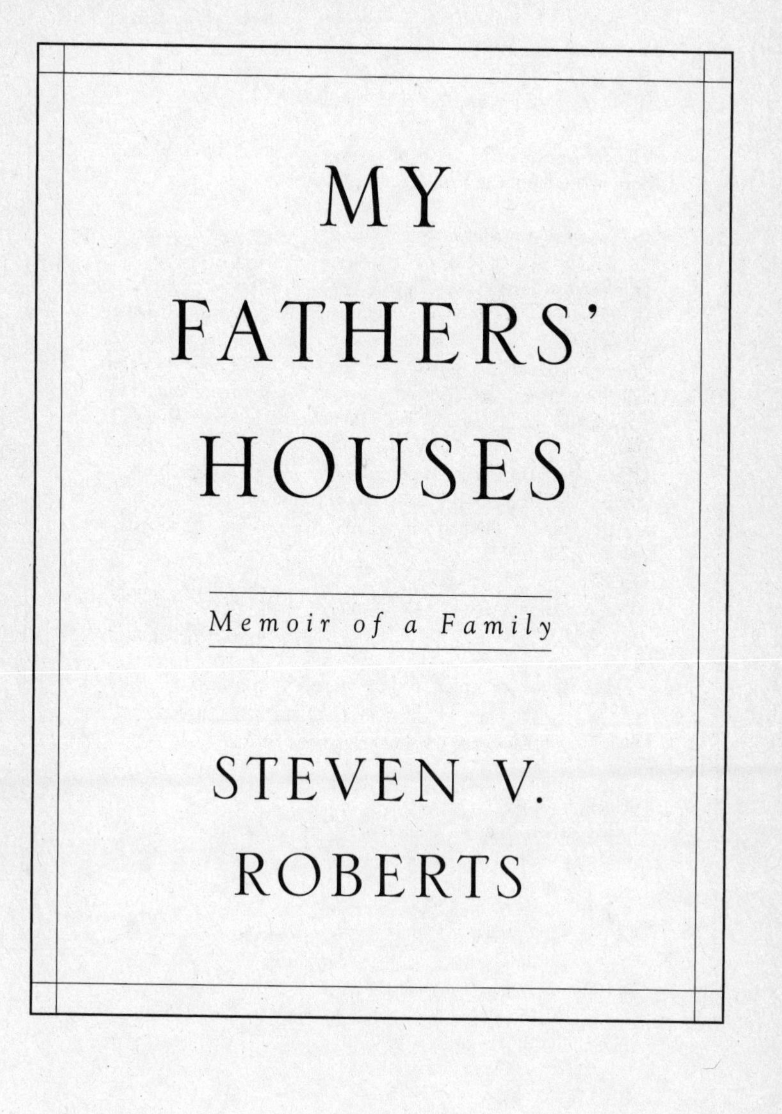

MY FATHERS' HOUSES

Memoir of a Family

STEVEN V. ROBERTS

HARPER PERENNIAL

NEW YORK • LONDON • TORONTO • SYDNEY

HARPER PERENNIAL

The writings of Lee Rogow are reprinted with the permission of Maggie and Zack Rogow.

All photographs are courtesy of the author.

A hardcover edition of this book was published in 2005 by William Morrow, an imprint of HarperCollins Publishers.

P.S.™ is a trademark of HarperCollins Publishers.

FIRST HARPER PERENNIAL EDITION PUBLISHED 2006.

Designed by Jennifer Ann Daddio

The Library of Congress has catalogued the hardcover edition as follows:

Roberts, Steven V.
My father's houses : memoir of a family / Steven V. Roberts.—1st ed.
p. cm.
ISBN 0-06-073993-2
1. Roberts, Steven V., 1943– 2. Jews—New Jersey—Bayonne—Biography. 3. Roberts, Steven V., 1943– —Family. 4. Roberts family. 5. Rogowski family. 6. Ethnic neighborhoods—New Jersey—Bayonne. 7. Journalists—United States—Biography. 8. Bayonne (N.J.)—Biography. I. Title.

F145.J5R63 2005
974.9'26004924—dc22

2004065473

ISBN-10: 0-06-073994-0 (pbk.)
ISBN-13: 978-0-06-073994-2 (pbk.)

06 07 08 09 10 ❖/RRD 10 9 8 7 6 5 4 3 2 1

To my grandparents, who started
the family story in America

MIRIAM AND ABE ROGOW

SADIE AND HARRY SCHANBAM

and to my grandchildren, who will keep it alive

REGAN BOGGS ROBERTS

JOHN ROBERTS HARTMAN

CLAIBORNE HILL HARTMAN

HALE McDONALD ROBERTS

ROLAND WILLIAM HARTMAN

CECILIA GRAY ROBERTS

CONTENTS

ACKNOWLEDGMENTS

My family has been wonderfully supportive throughout this entire project. My brothers Marc and Glenn and my sister Laura generously shared their memories and reflections. So did my aunt and uncle, Lee and Bernie Schanbam, and my grandfather's old friend, Julius Sorkowitz. Three cousins have done great work preserving the family history. Maggie Rogow collected and reprinted the writings of her father, Lee Rogow, my uncle Bussy. Jeff Mishlove captured many of our relatives on tape during a family reunion, and the memories of his mother, Rose Mishlove, my dad's only surviving sibling, were particularly helpful. Jeff's brother, Andy Mishlove, provided many of the photographs of our grandparents you see in the book. All my writing relatives, my partners in the family business, have always been an inspiration to me, but I feel a special connection with my cousin Zack Rogow, an extraordinary poet, teacher, and translator. Both of us, in our own ways, are carrying on the work his father started and never finished.

My children, Lee and Rebecca Roberts, and my daughter-in-law, Liza McDonald Roberts, read the manuscript and provided both affection and advice. My six grandchildren are all a constant source of delight and gratification, but a special word about the two babies, Roland William Hartman and Cecilia Gray Roberts. Roland got himself born two days after this book was finished. Cecilia arrived just as the paperback went to press. Good timing, kids.

A huge, heartfelt thanks to my mother, Dorothy Roberts, who carefully saved the letters she and Dad wrote to each other all those years ago. They gave me a priceless insight into their lives and times. Beyond that, she patiently answered my questions for hours on end, unearthed a trove of old photographs, and kept rooting me on. But then, she's been doing that for a long time.

I want to thank Joe Ryan, a dedicated historian of Bayonne, and Kathleen M. Middleton, author of two useful Bayonne books. Many old friends helped as well, particularly Barney Frank and Renee Cherow-O'Leary. My agent, Bob Barnett, provided his usual combination of warm friendship and wise counsel. My colleagues and students at George Washington University put up with my absences and distractions.

The folks at William Morrow are the best chums and cheerleaders an author could have: Kevin Callahan, Jennifer

Pooley, Sharyn Rosenblum, Samantha Hagerbaumer, Lisa Gallagher, Debbie Stier, and the boss, Michael Morrison.

Above all, there is Claire Wachtel, friend and editor for many years now. This memoir was her idea. She pushed me—gently at first, then more firmly—to take the risk. And she provided the framework and focus I needed to complete the project. No Claire, no book, it's that simple. My deepest gratitude.

Finally, there is Cokie. We've been together more than forty years, as partners and parents, lovers and friends. She hears my stories and heals my soul. She holds my hand and holds me up. And she still makes me smile. Thank you, angel, for everything.

MY FATHERS' HOUSES

A BOTTLE IN A BUCKET

I still dream about Bayonne. Usually I'm back living there, often in the house where I grew up, a two-family frame structure on a crowded block that ends at a low bluff overlooking Newark Bay. All the houses on The Block were the same, about twenty-five of them, separated by alleys so narrow that you always knew what your neighbors were arguing about or having for dinner. The Block was the center of my world for thirteen years, from my birth in 1943 until we moved all of five blocks away in 1956, and it could have been a European village, on the top of a mountain, surrounded by medieval stone walls. All the families knew each other, strangers were sparse, and you could walk to the shops around the corner for most of your daily needs.

That's no accident, I suppose, since most of the families, including mine, were only one generation removed from their Old World origins, and they re-created the patterns of life they had known in Poland and Russia, Ireland and Italy. There were a handful of Catholics on The Block, but most of the families were like us, Jewish people with roots in Eastern Europe—Lipkin and Lauton, Moritz and Hoch, Reznick and Levy. Some were manual workers, like my grandfather Harry Schanbam, a carpenter who had built the house we lived in with his own hands. Some in the next generation had gotten an education and become professionals. Yale Greenspoon's father taught at the high school, Artie Schackman's dad was a photographer. Many owned small businesses. The Penners ran a furniture store on Broadway where Mom bought our baby supplies. The parents and grandparents of the girl I took to the junior prom ran a hardware store. New York was only a short bus ride away, but "the city," as we called it, seldom intruded into our lives. Broadway and Forty-second Street in Bayonne (there really is such an intersection) was light years away from the more famous corner just across the Hudson River. My father commuted daily to "the city," where he ran a small children's book publishing company, but few if any of my friends had parents who did that. Most people lived and worked, met and married, grew old and died, all within the confines of this urban village. Bayonne was not exactly Anatevka, and we didn't have any fiddlers on our roofs, although we did have Mr. Friedberg, who delivered seltzer to the door in blue glass bottles with silver spritzers. But when I saw the movie *Avalon*, Barry Levinson's ode to the Jewish community of Baltimore, I felt a pang of recognition. In that movie the immigrant generation clings to the old neighborhood and the old ways, and when their kids move to the suburbs, the old folks find the adjustment disorienting. Bayonne, like Baltimore, was actually closer to the Old Country than the suburbs were to the inner city.

Bayonne is a peninsula, about five square miles, surrounded on three sides by water: Newark Bay to the west, the Kill Van Kull on the south, and the Hudson River on the east. In fact, after we left The Block, I could catch a glimpse of the Statue of Liberty from my new bedroom window. But I've never been there and I'm not sure why. I guess you don't play tourist in your hometown. During my childhood, you could enter and leave Bayonne in only two ways—by city street to Jersey City and by bridge to Staten Island—so the word "insular" really did apply. I flew over it recently, heading for Manhattan, and I was struck again by how distinctive Bayonne is. You can pick it out immediately from the air. And since it was such a separate and self-contained place, it had a strong sense of identity. One public high school, one daily newspaper, one downtown shopping district. To this day I meet people all over the country who want to tell me about their connections to Bayonne. My friend Barney Frank, now a congressman from Massachusetts, who grew up there, says people always talk about being from Bayonne because they are "so proud of rising above their humble beginnings." But I don't think that's quite right. I think it's because Bayonne is a real place, with a long history, dating back to its discovery by Dutch explorers in the seventeenth century. It's not a fake city, bordered by arbitrary lines on a suburban map and bearing some insipid variation of the name Parkforestglenwood.

It's also true that Bayonne has become something of a joke, like Secaucus, employed as a punch line by comedians and cartoonists. One of my favorite references is a *New Yorker* cartoon showing a man sitting at a bar and saying to no one in particular: "I'm a citizen of the world, but I make my base in Bayonne." Jackie Gleason once did stand-up comedy at the Hi-Hat Club in Bayonne, and his TV show *The Honeymooners* was loaded with local references. If he frequently threatened to send his wife, Alice, "to the moon," he often vowed to dispatch his pal Norton to Bayonne. My brother Marc remembers Gleason portraying a pitchman in a TV comedy skit. If you call in right away, he promises, and order the food chopper or vacuum cleaner he's selling, he'll throw in a free pennant from Bayonne Technical High School. Who could refuse that offer? The *New York Times* obituary of the comic Rodney Dangerfield noted that he got his start playing "dingy joints" in places like Bayonne. As Dangerfield himself might have said, my hometown "gets no respect." A Navy ship was once named for the city, the USS *Bayonne*, but in the middle of World War II it was actually given to the Russians, who then gave it back. "How pathetic," says Joe Ryan, a Bayonne historian who told me the story. In the movie *How to Lose a Guy in 10 Days*, Matthew McConaughey takes Kate Hudson home to Staten Island to meet his decidedly working-class parents, who are constantly playing a card game called Bullshit. Matt tells Kate the walls are so thin in their house that when they flush the toilet, people "can hear it in Bayonne." I sat there in the darkened multiplex thinking, "There it is again. Bayonne follows me everywhere."

I had the same reaction when I read a story in the *New York Times* describing how scenes from Steven Spielberg's blockbuster *The War of the Worlds* would be filmed in Bayonne. That's typical. My hometown finally gets noticed by Hollywood and it's invaded by Martians! In fact the star of the movie, Tom Cruise, lives in a house not far from where my grandfather once ran a carnival and promoted boxing matches. The *Times* writer couldn't help taking a couple of cheap shots, calling Bayonne "downtrodden" and "one of New Jersey's backwaters." But Cruise got it right. After buying an espresso with steamed milk at Chez Marie, a coffee shop on East Twenty-second Street, he told the owner: "This is a really nice town."

Bayonne was actually part of movie folklore before Spielberg discovered it. As a boy I used to play basketball at the PAL, the Police Athletic League. The court was on the second floor, and many nights I would pass a ground-floor gym where local boxers trained. One of the fighters I would see occasionally was Chuck Wepner, known with some derision as The Bayonne Bleeder during his inglorious career. Chuck was working on a beer truck when he got tapped, out of the blue, to fight Muhammad Ali, during Ali's championship reign. Wepner took the champ to the limit, before losing, and Sylvester Stallone, who saw the fight, recounts his reaction.

"That night I went home and had the beginning of my character," Stallone writes in *The Official Rocky Scrapbook*. "I was going to make a creation called Rocky Balboa, a man from the streets, a walking cliche of sorts, the all-American tragedy, a man who didn't have much mentality but had incredible emotion and patriotism and spirituality and good nature even though nature had not been good to him." The screen version of Rocky trained for his fights by running up the steps of the art museum in Philadelphia, but the real Rocky, Chuck Wepner, ran through the streets of Bayonne.

Whenever any of us drive home, we look for landmarks telling us the journey is almost over. For some it's a river, or a park, or a tall building. Where I live now, outside of Washington, D.C., a Mormon temple hovering over the Beltway tells me: only a few more minutes. As you drive toward Bayonne, one sure sign of home is the network of rusting overpasses and train trestles that criss-cross the Meadowlands outside Newark, decaying remnants of an industrial age that's largely gone today. (That same landscape has been immortalized by the opening scenes of *The Sopranos*, which is filmed in North Jersey, and some episodes include shots of Bayonne.) But the most obvious signals that the trip is ending are the bright green signs on the New Jersey Turnpike, which announce, BAYONNE, EXIT 14A. For many Americans, if they know anything at all about Bayonne, it's that one fact: exit 14A on the turnpike. Few drivers ever take that exit, which opened in 1956, as they speed by to New York City or New

England or Philadelphia. The readers of this book will get off the turnpike at 14A, but without paying a toll.

In one sense Bayonne prepared me well for a larger life and a larger world. I knew who I was and where I was from. I was connected by innumerable cords to people and places that gave me strength and identity: my parents, who both grew up in Bayonne and met on my mom's seventeenth birthday; my three living grandparents, who all lived within four blocks; Mr. Levine, who ran a small fish and vegetable store around the corner and wore the same gray sweater for twenty years; my mother's cousin Davie Feldman, who was our milkman; my mother's best friend from childhood, Charlotte Simon, who lived only one house away. My parents and I attended classes in the same building, although it was a high school in their day and an elementary school in mine. On The Block I was safe, secure, loved. I even had a number, 174, the address of our house, but the number wasn't a badge of anonymity. To the contrary, it marked my place, where I belonged. It was so much a part of me that it could have been tattooed on my wrist. The same was true of my phone number, Bayonne 3-1409. One problem, however: The number for the kosher chicken market was almost the same, and we would get frantic calls on Fridays from Jewish housewives, caught without a centerpiece for their Sabbath meal.

But in another sense, Bayonne didn't prepare me at all for life beyond The Block, and one of the first people to teach me that was Lisa. The summer before I left for Harvard I was working as a messenger at *Newsweek* magazine in New York, and every day I would take a train into the city and connect to a subway that left me off at Fifty-third and Madison. I started noticing this cute girl who seemed to be on the same schedule, since we passed each other many mornings on Madison Avenue, walking in opposite directions. But because I was a social misfit, the flirtation took weeks. First we smiled, then we waved, finally we stopped to chat. We were exactly the same age, it turned out, both aspiring journalists, both working at magazines, both headed for New England colleges in the fall.

But there the similarities ended. She was a wealthy, well-bred Protestant from suburban Connecticut, one of whose ancestors had been secretary of state under Woodrow Wilson while my grandparents were still learning English and working on dairy farms in Bayonne. On one of our first dates we went to a small restaurant near our offices and I ordered Chianti. Since Chianti bottles then were rounded on the bottom and couldn't stand on their own, they came sheathed in wicker bases. It was the one wine I recognized, and that was only because pizza places in Bayonne used to decorate their checkered vinyl tablecloths with candles stuck in those quaint containers. For some reason I asked that the wine be chilled. The waiter looked at me like I was nuts. You don't chill red wine, he sniffed, and besides, what would he do with the wicker base? But I was in too deep to back off. Chilled, I said, we want it chilled. So the waiter stripped off the wicker with a knife, and plopped the pale green bottle into a bucket of melting ice. I can still see that bottle, looking undressed and unstable, sloshing around in the bucket. That bottle was me. Bayonne was my wicker basket. But now it was gone, and I had to find a new way to stand on my own.

This was the summer of 1960, the Eisenhower era was ending and the Kennedy era was about to begin. The fifties were over, the sixties were dawning. Lisa and I didn't realize, as we sat there sipping our well-chilled wine, that we had been shaped by a world and an era that was rapidly disappearing. In just a few years, girls would be taking off their girdles—and their bras. We'd be listening to Carole King and The Beatles, not Patti Page and the Kingston Trio. The "pill" would become as popular, and as available, as pizza. First for the *Harvard Crimson*, and then for the *New York Times*, I would find myself covering protest marches, not pep rallies, getting drenched in tear gas, not beer. It was the age of Sex, Drugs, and Rock 'n' Roll, and by 1970, when I went back to my tenth high school reunion in Bayonne, I wrote a piece for *The New York Times Magazine* about my classmates called "Old-Fashioned at 27." The confusion we all felt was summed up by my next-door neighbor on Thirty-first Street, Joyce Lauton Nestle, then a teacher at the high school: "We're scared whether we're going to make it. We believe in the new ideas but we don't know whether we can live with them. Like the idea of property. I like the idea of a commune, but I'm used to the old way. I don't think we're flexible enough for the changes that are coming, and it hurts so much to be insecure." It sure did hurt. We were all "used to the old way." But if I was able to "make it," to cope with the changes cascading around me, to find a way to stand on my own, it was because of Bayonne and the people who loved me and shaped me in those years, before I got off that subway at Madison and Fifty-third and noticed a girl walking down the street.

The story really starts with my paternal grandfather, Abe Rogow, born in the Bialystok region of what's now eastern Poland in 1891. He was a man of fierce determination and unquenchable energy, who made a new life for himself not once but twice in strange lands. As a teenager he embraced the Zionist ideals of Theodor Herzl and decided to become a *chalutz*, a pioneer in Palestine. Later he immigrated to America, married his sweetheart from the Old Country, and raised four children in the New World.

He was a man of words and ideas, and his passion for both pulsed through the family and left its mark on all of us. When it came to making money, he was, to put it bluntly, a con artist. His first fortune—he made and lost several of them over time—came from running carnivals and promoting prizefights. But the gambling was always a sideshow to the main event in his world, the life of the mind. I grew up thinking the family business was writing and politics, and that tradition went back to Europe, where my grandfather's brother Marcus was a newspaper editor and his sisters were early and active Bolsheviks. On weekday mornings Abe didn't say prayers, like observant Jews do, he read the *New York Times*. On weekends he didn't go to synagogue, he watched TV shows like *Omnibus* and *Meet the Press*. These were his religious devotions. The *Times* was his scripture. Lawrence Spivack, the moderator of *Meet the Press*, was his rabbi. Abe was a man of many opinions and few doubts. He once published a booklet outlining his own plan for world peace and challenged Winston Churchill to a debate, a challenge that was actually answered by Churchill's son Randolph. The debate never came off, but it reinforced the conviction in Abe's descendants that the world was dying to hear every word and thought produced by a Rogow. All three of Abe's sons, including my dad, worked with words most of their lives. All three of my siblings are in the family business as well: My brother Marc teaches at the Harvard School of Public Health and writes widely on health policy; my brother Glenn, after many years in journalism and politics, now runs trade associations and devises their legislative strategy; my sister Laura advises museums on educational programs, teaches the management of nonprofit organizations at several colleges, and just signed her first book contract. If you include my siblings, my cousins, my children, and their spouses, I can count about a dozen of us who have published books of one kind or another, from sexy detective novels and volumes of love poetry to an encyclopedia of parapsychology and a professorial tome on the cancer risks of toxic chemicals. In fact, two of my writing relatives, novelist Miriam Moore and journalist Miriam Horn, are named for my grandmother, Abe's wife, Miriam.

Abe wanted me to be a writer, to be an American, to carry on the family business, and he couldn't contain his pride, or his quirky sense of humor, when I went to work for the *New York Times* in June 1964 as a research assistant to James Reston, the Washington bureau chief. Eight months later I got a letter from Abe, sent from Bayonne, that began this way: "Dear Steve. This is Grandpa Abe. I am setting [sic] near the den window watching the storm raging on Newark Bay. The first time in 18 years since I built this house, it is bitter cold in the house. I can't blame the landlord, because I am the landlord. Talking about storms. I just got a brainstorm. I called Cortshire [a tailor in New York where he could get a discount] and I asked them to send some swatches to you. I want to blow you to another suit of clothes. I want you to be the best dressed on the Times [a hopeless ambition]. I know you will be up against some smart dandy's [sic]. I'll bet Russell Baker [then just starting to write a column for the *Times*] has a pair of cufflinks for every wisecrack. And Mr. Reston a tie pin for every mood."

An even bigger influence was my father, Will, Abe and Miriam's firstborn. At NYU during the thirties, he founded a magazine, *The Square*, and the few faded copies that survive pass for precious family heirlooms. A few years later, in the middle of World War II, he started his own publishing business, but money was scarce so he wrote many of the books himself. Since his name was already on the company, he needed a nom de plume, and he picked Jeffrey Victor, a combination of his twin sons' middle names. (My name, Victor, reflected my parents' hopes for the outcome of the war; they always told us that my brother Marc's middle name, Jeffrey, meant peace. Dad also wanted us to have strong middle initials, in case we wound up as writers with our own bylines. Good thinking, Dad.) So in a sense Dad gave me my first byline, when I was still a toddler, and as I write, nine of the ten books he wrote are arrayed on the bookshelves behind me, reminding me of his legacy.

But I don't need much reminding. From the time I wrote a composition in fourth grade about Yellowstone Park—heavy with flowery adjectives, but showing a small glimmer of promise—he encouraged and counseled and guided me with an editor's eye and a father's heart. Like all kids, I soaked up praise like a dry sponge, and doing something well that my father respected was a double bonus. He was my first and best teacher. For every word I ever wrote, I looked forward to his reaction. For every TV show I ever appeared on, I called my parents to tell them to watch. One evening in May 1997, after I made one of those calls, my mother blurted into the phone, "Steven, your father has had a stroke." My wife, Cokie, and I rushed up to New Jersey the next day to see him. He was talkative in the hospital, asking about the baseball scores and the stock market. Later that night, he died from another stroke. I miss him deeply. I want to hear his voice on the phone, saying that one of my articles in the paper or comments on TV was "great, just great." I want to tell him that I'm still in the family business, that I'm still living in the house that my fathers built for me.

Dad was so enthusiastic about my writing career in part because his own was so brief. His years as a children's book author ended shortly after the war did, and he spent most of his working life as a businessman, not an intellectual. He never made his mark as a writer, and his hopes and frustrations created a powerful message that poured into me from a very early age. Writing is what you do. It's what we do. "It was in your blood," says my mother, "it was just destiny."

I sensed all this at the time, but I understand it even better now. In 1939, while my father was trying and failing to make his fortune in Nevada's nascent gambling industry, my mother wrote him a letter that I discovered only recently. "Will, dearest," she wrote, "since I read your letter this morning the conclusion has been growing on me that your opportunity is staring you in the face. You know it's there, you've been conscious of it for years, but always as something vaguely future. Will—it's not! It's for you to do right now. All this circumlocution is in reference to your writing. Will—for God's sake—for your sake, my sake—for the sake of the unborn Rogow progeny—begin to write!" Her advice gets even more pointed: "Leave fiction entirely untouched for awhile—I believe your ability in that sphere of writing is underdeveloped whereas your critical faculties are tremendously keen and unquestionably of importance." She adds: "The Nation and all those other magazines of that type want the sort of stuff you can give them—that you are prepared to give them right now."

She turned out to be wrong. Dad never wrote for *The Nation* or any of "those other magazines." But twenty-two years after she wrote that letter, their firstborn child (I beat my twin brother by ten minutes) published his first professional article. In *The Nation*. So maybe Mom was right after all. Her advice just skipped a generation. And now that I've read that letter, long buried in a box in the basement, I have a better idea of what Dad was feeling in the summer of 1961, when I was eighteen, and the article I had written about a student political meeting was accepted for publication. I was home in Bayonne the day the piece appeared, and Dad insisted that we get in the car and drive into New York and find it. It took a while, not many newsstands carried *The Nation*, even near the magazine's editorial offices in Lower Manhattan, but when we finally found the new issue, with my name right there on the cover, he was more excited than I was and bought every copy. After all, he had put my name on the cover of his own books many years before. And he had never seen his own name on the cover of *The Nation*. As supportive as he was, however, Dad also taught me lessons in humility, some deliberate and some not. The day I got my first byline as a *New York Times* staff writer I called him up with great excitement. The byline had been a surprise, a sign that I had passed my trial period and was now a full-fledged reporter. He was pleased but confused. He'd read the *Times* that morning. What was the story about? When I told him, a little piece about *Ebony Magazine*, he said: "I read that piece. I didn't notice your byline." So much for fame.

After Dad drifted out of book publishing, he made a lot of money in other business ventures and provided very well for his family. One day, late in his career, he sat in a restaurant, looking at the menu, and said with considerable pride, "I can have the Businessman's Special. I'm a businessman!" But there was a hole in his life, a hole of unfulfilled ambition, and as I reread old letters I can feel his frustration poking through that hole. He loved the fact that I was a writer, but he harbored a tiny shred of regret that he had not been one as well. During my junior year at college, he wrote a letter to the *Crimson*, the Harvard daily, commenting on two of my articles, and I have the note he enclosed to my friend Bruce Paisner, then the paper's editor. "I can't recall when I had so much fun writing something," he wrote to Bruce. "I hope you can find some use for it. Needless to say, if it does appear in the Crimson, I will become absolutely unbearable with my friends." The edge of anxiety in his note is unmistakable. He wanted to be published. He wanted to be noticed. (The *Crimson* did use the letter, much to his delight.) Most of the time, the fact that he was sending us to college and glorying in our accomplishments was enough for him. But occasionally he wanted more, a public recognition of his own abilities, and that didn't happen often.

In later years, when I appeared on *Meet the Press*, I told Tim Russert about the role the show had played in Grandpa Abe's household during my childhood. In a gracious gesture, Tim asked for my dad's address and sent him a note, enclosing a *Meet the Press* baseball cap. Next time Steve appears, he added, I'll send you a T-shirt. Dad was pleased but asked me, only half in jest: Why doesn't Russert put me on his show and send you a T-shirt? At one point during my college years, I briefly raised the possibility of getting involved in local politics. In a letter he wrote to Marc and me, Dad warned us off the "venal" politics of Hudson County, but added wistfully: "I would like to see one of the three of us in Congress."

The year after he wrote to the *Crimson*, Dad sent me a letter saying, "I am hoping to be able to get back to my writing. I haven't written a line in about a year." His aim, he said, was not to write something "profoundly meaningful." Rather, "I am hoping that it will be funny, or at least amusing." But in a burst of self-knowledge he went on to say, "I have every quality of a writer except one. I just lack that self-discipline that impels one to sit down at a typewriter and write." He was right. A later note mentions a manuscript of twenty pages or so and then the references to his writing disappear. I have a sheaf of half-started, never-finished efforts. In another letter he pointed out that his four children were fortunate, because we inherited from our mother qualities of persistence and tenacity that he never had, at least when it came to writing. "The children of Will Roberts will conceive of great books, ideas to move the minds and hearts of men," he once wrote in a sentimental Mother's Day tribute to Mom. "But the children of Dorothy Roberts will write the books, set down the ideas, and men will be moved."

A scalding tragedy shadowed Dad's life and deeply influenced mine: the early death of his younger brother. He was named Leon, but his byline was Lee Rogow, and he was a scintillating figure who wrote reviews and short stories and plays and seemed headed for a brilliant literary career. He was killed in a plane crash in 1955, at age thirty-six, while making a training film for the U.S. Army. Bussy, as the family called him, was many things Dad was not—debonair, sophisticated, a regular at New York opening nights and nightspots—and I think my father lived vicariously through Bussy's adventures. I was twelve when my uncle died, and I always felt, keenly and clearly, that my father then transferred many of his fondest hopes for Bussy onto me. When I was twenty-one, and enmeshed in a long and angry confrontation with my father about the woman I was dating (and have now been married to for almost forty years), I wrote him a letter, trying to explain who I was and what I stood for. "I am my father's son, and the son of his fathers," I told him. "I find great strength in knowing that by writing I am doing the work that you wanted to do but never could, and that Uncle Bussy began and never finished." I felt so strongly about that legacy that we named our first son Lee, in honor of my uncle. Bussy was so vibrant, and his death so traumatic, that other family members also saw him as an inspiration. His son Zack, a noted poet and translator, says, "It was my task to have the career he never had." And his late brother Murray once confessed to an "insane, infantile desire to be my brother Lee."

In a way, I see now, Dad's disappointments made my life easier. I had all of his affection and encouragement, but he never produced a body of work—or a public reputation—that I had to live up to. I went to college with many children of famous people, and while their parents opened a lot of doors for them, they also had to struggle, and not always successfully, to establish their own identities. I was free of that burden. No one knew Will or Bussy Rogow or compared me to them. I could join the family business but be myself at the same time. My ancestors fed and watered my roots, but they didn't block out the sun or stunt my growth.

There is a bright thread that runs through the pattern of my life, a dream that was born in Russia and Palestine and flourished in America, that was passed on to me from Grandpa Abe and Uncle Bussy and my father, Will. I realized this one evening when I was walking through a parking lot after dinner and watching my father from behind. The whole set of his body and the rhythm of his gait reminded me of Abe. When I shared this observation with my brother Glenn, he laughed. I was watching you from behind, he said, and you reminded me of Dad! For years, when I passed a mirror or a plate glass window, and caught a glimpse of my reflection, I'd think of my twin brother, Marc, and say to myself, "What's he doing here?" Now, with my hair more gray and my neck less taut, when I see my own image, it's my father looking back at me, more than my brother. I even have the same small growth on the right side of my nose that Dad did. Many people have influenced me deeply—my mother, my wife, my children, my siblings, my first boss at the *New York Times*, Scotty Reston. But the fathers who walk like me and look like me play a special role, keeping me strong and keeping me straight.

In one of my dreams about Bayonne, I found myself in my grandparents' house, the one overlooking Newark Bay, where

Abe sat by the window and wrote out his offer of a new suit of clothes. I looked around and saw my grandfather sitting in a corner—a big shock, since he's been dead more than twenty-five years—but I went over and embraced him. He asked me about my latest book, and I told him he should read it, that there were many stories about him in it. So here it is, Pop. A story of a town and a time and a boy who grew up there.

OLD WORLD ORIGINS

Bialystok, like Bayonne, seems to inspire comedians and scriptwriters. The main character in the hit musical *The Producers* is Max Bialystok, a blustery buffoon who deliberately tries to lose money by putting on a show called *Springtime for Hitler.* (Then there are bialys, a form of onion roll popular in New York delis that supposedly originated in Bialystok. But I never saw one when we visited that city.) When my grandfather was born there as Abraham Rogowsky, on the cusp of a new century, Bialystok was a prosperous textile center. It's now in eastern Poland, but then it was part of the Russian sphere. My grandparents always thought of themselves as Russian, not Polish—I think being Russian was a status thing for them, Poles were peasants in their view—but their pride came with a price. If you were Russian, the czar was a distant but disturbing presence, and

I'm reminded of the song from *Fiddler on the Roof*, where the villagers voice the fervent prayer, "God bless and keep the czar—far away from us!"

Avram, as he was known in Yiddish, had three older sisters and an older brother. He occasionally used the middle name Bear (or Baer), but it was not part of his official name. When he was a baby, and fell sick, his mother called him Bear to fool the evil spirits. Abe's father was a commission merchant, a middleman who sold the textiles made in the area to tailors and dressmakers. "He was a scholarly man," my father once wrote, "and often expressed a wish that he could own a bookstore, so that he could sit all day and read books." The elder Rogowsky never achieved that ambition, but he had the means and the motivation to educate his children, including his daughters, a revolutionary idea at the time. In fact, teaching girls caused such a scandal that one friend asked my great-grandfather if he'd consulted the rabbi on the subject. Yes, came the reply, and the rabbi said don't do it. But the Rogowskys were a stubborn tribe, particularly when it came to clerical authority.

My great-aunts went to college in Switzerland where, legend has it, they met Lenin, who was then in exile. The experience left them with the credentials to teach French and a lifelong commitment to Bolshevism. Their brother, Marcus, became a teacher and writer, working for a time as a tutor to the children of Russian army officers. Political radicalism ran strong in the Rogowsky blood and Dad wrote that Marcus helped foment some of the revolts that plagued the Russian army around 1905. He was a Menshevik, however, a more moderate Marxist faction than the Bolsheviks, and some years after the army uprisings he came to America as a newspaper correspondent. When the czar fell in 1917 and the Kerensky government, which followed Menshevik principles, briefly took power, Marcus was summoned back to Russia to join the revolution. According to one story, his way was paid by a Menshevik sympathizer in Russia's Washington embassy. With war raging across Europe, the only way he could reach Moscow was from the east, so he traveled by ship through the Panama Canal and then to the Japanese port of Osaka. From there he set sail for Vladivostok, Russia's gateway to the Pacific. Abe received a postcard, mailed from Osaka, and then silence. He never heard from his brother again.

After leaving for America, Abe had totally lost touch with his sisters, who disappeared behind the Iron Curtain. More than fifty years later a cousin immigrated to Israel, tracked Abe down in Bayonne, and told him that his sisters were still alive. Eventually he visited them in Moscow, and learned what happened to Marcus: their brother did make it back home, just in time for the Kerensky government to fall to the Bolsheviks. He survived for a few years, working as an editor at *Pravda*, the principal paper of the Communist Party, but was liquidated in the purges of 1923. That certainly makes me one of the few *New York Times* reporters with a great-uncle who worked for *Pravda*, and I was always jealous of my brother Marc, who was named for this dashing figure in family history. Why did my twin, the college professor, get that richly resonant name instead of me, the newspaper writer?

As a boy, one of Abe's jobs was delivering to the railway station bolts of cloth that his father was sending to merchants around the country. He often spoke of how his sisters, the Bolsheviks, would give him packages to ship at the same time. They looked, on the outside, like another bundle of textiles. But in reality, those packages contained the *Iskra*, or "Spark," the Leninist newspaper that the Bolsheviks were smuggling into Russia. Many years later Abe was still furious at the deception. Imagine, he used to say, if I had dropped one of those packages and it had split open. I would have been shot on the spot.

As the youngest, Abe was close to his mother, Rose, who worked in the textile business and also had to play host to the buyers who came through town and often stayed at the Rogowskys'. After Rose died, when Abe was about twelve, his father married a widow with several of her own children, and the new wife brought along an unmarried sister as well. Abe and the sister hated each other, and as Dad recalled, "To the end of his days, my father could get really worked up talking about the step-mother's sister, who, he said, always favored the step-mother's children, giving them the best food, clothes, etc." I also have a picture of Abe as a young teen, posing with classmates at some sort of military school, and I can't imagine that kind of regimentation helped his frame of mind. He was a rebel to the core.

The picture of the boy foreshadows the man. Abe was al-

Grandpa Abe with his military school classmates in Bialystok, then part of Russia, before he left for Palestine in 1907. He's in the top row, second from right.

ways thin, even gaunt, perhaps because he was an early health food faddist and seemed to subsist, for long periods, on figs and wheat germ. But his wiry frame always contained a coiled intensity. It's no surprise to me that Abe could get "worked up" over the memory of those early years, because in my experience he always seemed to be "worked up" over something, from the employees who were stealing his cash, to the politicians who were spurning his counsel.

If the Rogowsky household was in an uproar during Abe's teenage years, so was the town of Bialystok. A major pogrom in 1906 killed dozens of Jews, and I have to believe that the anti-Semitism of the time helped convince Abe to leave. In that era the Jews of Eastern Europe lived in a constant state of fear, without civil rights or political power to defend themselves. Some reacted by retreating into their religion and walling off the outside world as best they could. I love the book *In My Father's Court* by the Yiddish writer I. B. Singer, who describes how the Jews of Poland set up their own private legal system, presided over by the rabbis, to adjudicate small disputes within their own community. Others turned to politics as their salvation, often embracing Marxism and its doctrine of a revolutionary proletariat. For Abe the answer was Zionism, an idea that swept through the shtetls of Eastern Europe and encouraged many young men to emigrate to Palestine, then a part of the Ottoman Empire, and help build a Jewish state in the desert. My grandfather seized on the Zionist message with the ardent—and often arrogant—zeal that marked his entire life. This was not a man of half-measured responses or halfhearted convictions. He never lost his Old World accent, and I have sharp memories of Abe pointing his finger for emphasis and insisting, "Think of that," although it always came out closer to "tink a dot." (He also never learned to say an English *v* properly, so when he counseled his grandchildren to eat their vegetables, the word came out "wedge-a-tables.") Dad said Abe left Bialystok with the approval of his family; Marc says he was sent to collect a debt for his father and stole the money to finance his trip. Either way, I know this: In 1907, sixteen-year-old Abraham Rogowsky made his way to Palestine.

Near the end of the nineteenth century, Baron Rothschild and other Zionists had started forming and subsidizing agricultural communes among the Jewish settlers, and Abe apparently joined one. Sixty-two years later he wrote to me with his memories of those early days: "When I first arrived in Palestine at the age of 16, the surroundings impressed me so much that I was enchanted. I was a watchman on a vineyard in Rehoboth. I could hardly wait for nightfall. Of course we had a sky and stars in Byalistok [sic] too. But not like in Palestine. The clouds floating overhead were as though they were telling you something. Every minute every night something different. I could not tear my eyes away. I felt like a communion with the world above."

Jews were pouring into Palestine, many through the port of Jaffa, an old walled city with narrow streets and limited space. Some enterprising Englishmen decided to build housing for the new immigrants and develop a city on the sand dunes outside the walls. They cut through one street, called Herzlia after the Zionist prophet, and quickly sold the lots. When they were ready to build a second street, named Binyamin after a wealthy Zionist benefactor, they hired Abe and his friends in the commune to do the work. Abe used to describe the stuffy Brits in their bowler hats and umbrellas, and how the young workers would stand behind them, mimicking their motions as the land developers showed prospective buyers through the area, pointing with their "bumbershoots" to where new schools and shops would be built. I have a picture from that time showing five young men, all deeply tanned, wearing hats and long-sleeved white pullovers, leaning casually on shovels and peering intently into the future. They're grouped around a large cart filled with dirt and marked with a Jewish star and the number 330. Behind them, a few half-finished buildings are visible. They radiate an air of cocky self-assurance, but even those brash young men had no idea that they were building the second street in a city that would become a world capital known as Tel Aviv.

Abe loved Palestine and felt an easy kinship with the local Arabs. When he was particularly tan, he used to say, he was often mistaken for one. In later years, of course, as immigration continued to increase, so did tensions between Arabs and Jews, but Abe lived there during a brief time when the two populations shared the land in relative peace. But he couldn't stay. Back in Russia he was subject to the draft, and if you didn't show up—particularly if you were Jewish—your

Abe with his kibbutznik buddies working on the second road ever built in Tel Aviv. He's third from the right, behind the dirt-filled cart.

family could suffer severe penalties. So he returned home and was greeted as a hero. The young people of Bialystok would gather in the evening on the outskirts of town, to build bonfires, sing songs, and tell stories, and this dramatic figure, with his tales of life in a new land, commanded center stage. But during the day he didn't have much to do, and in his wanderings around town he would pass a photography studio. There, displayed in the window, was a picture of a young woman, fiery and bright-eyed. He fell in love with the picture, and one evening, at one of those gatherings outside of town, he met her. "You're the girl in Warshawsky's window!" he exclaimed, and indeed she was. Her name was Miriam Wasilsky.

Miriam grew up in a small town called Eishishok, not far from Bialystok. She was one of six children, with three brothers and two sisters, and her father was a poor man, scraping out a living as a bookbinder and monument carver among other jobs. But there was a photographer in town, and the Wasilskys had enough money to pose for a family portrait in which Miriam appears to be about seventeen. Her father is an austere figure, with a dark beard and skullcap, and her mother seems formal and formidable, in a high-necked, long-sleeved dress that reaches the floor. Not a lot of warmth there. Her brothers all wear flat, round-brimmed hats and double-breasted jackets—one has his hand tucked into the folds of the heavy material—and her two sisters wear dresses with lace collars. I have trouble thinking of this family group as my ancestors, everyone seems so awkward and uneasy. Everyone except Miriam. She stares into the camera boldly, almost insolently, with a slight half-smile. If Abe was spindly, Miriam was solid. If he was crazy, she was calm. If he was a character, she was a charmer. Even now, almost a century later, that photo of her takes my breath away. No wonder that Abe, on his deathbed, would hold the same picture, tears streaming down his face, and murmur, "She was so beautiful."

Beautiful, strong, and savvy. Miriam's mother was sickly, and she had a lot of responsibilities at home. She used to joke that if she ever went out with friends, she had to take a few siblings along. Eager to escape her small town, she moved to the nearest city and found work in a yard goods store, selling cloth that seamstresses would then make into

Grandma Miriam, second from the left in the back row, with her parents and siblings. This was probably taken in their home village of Eishishok, now in Lithuania, in about 1910.

dresses. Used to hard work back home, she could tear open the wooden crates containing the material with her bare hands. And she was a natural saleswoman. She could take a bolt of cloth, drape it over her ample frame, and convince the customer "that she would be Queen of the Ball," as Dad put it. "Everything looked good on her," adds Miriam's only daughter, my aunt Rose, named for Abe's mother. This salesmanship turned out to be a very useful trait. Often during their long marriage, when one of Abe's harebrained schemes

1912 December 8, Bialstok, Saturday evening
Take this remembrance with you, so that when I'm not near you will know that you once had a true close friend.
Your friend Miriam

Abe and Miriam, in Bialystok in 1912, before they were married. Her handwritten Yiddish inscription is below the picture, with the English translation above.

didn't work out, Miriam kept the family afloat with her business sense.

Soon the young couple agreed to be married, but there was one problem: the czar's army. The rule was that once you went in, you were the army's problem, not your parents'. So they devised a plan. Abe would allow himself to be drafted, escape as soon as he could, make his way back to Palestine, and send for Miriam. I have a picture taken of the two of them at just this point in their lives. It's in profile, and they're gazing off to the right at something I can't see. Their future grandchildren, perhaps? Beneath it Miriam has written a message to Abe in Yiddish, and here's the English translation: "1912 December 8, Bialstok, Saturday evening. Take this remembrance with you, so that when I'm not near you will know that you once had a true close friend. Your friend Miriam."

As a boy, I always heard this version of Abe's escape: He put on civilian clothes under his army uniform, and wore a specially made cap that looked like military headgear on one side. But when you flipped it over, it became a civilian cap. In another version, he was only issued boots by the military, not a full uniform, but was told to tuck his pants inside the boots to mark him as a soldier. Both versions agree that as soon as the troop train stopped to take on water, Abe jumped off. Perhaps he shed his uniform and flipped over his cap; or maybe he just untucked his pants from his boots. Either way, he was a fugitive, a lovesick fugitive, heading for Palestine to make a new life for a new wife.

Somehow Abe got to Odessa, a port on the Black Sea where ships departed for the Middle East. He had no papers, so he couldn't travel legally. But through the Zionist underground network, he met a family with twelve children, the youngest of whom had just died. So he posed as the oldest child in the family, and everyone else moved down a notch. Of course the genders didn't match anymore, but as Abe used to tell the story, the customs inspector counted knees as the kids sat on a bench. Since the family had papers for twenty-four knees, they were waved through. I imagine some cash changed hands as well, causing the inspector to have a momentary bout of blindness. It wouldn't be the last time my grandfather greased a bureaucratic wheel with a well-placed payoff.

By the time Abe got back to Palestine, however, the land was in turmoil. Ottoman rule was collapsing, World War I was approaching, and it was no place to bring a bride. Abe would write from different places, saying he had found the right spot, then change his mind. One story has it that he finally wrote to Miriam and said, "Change of plan, meet me in Brooklyn." Miriam had a brother, Morris, who had already moved to New York and established himself in the business of making headstones for cemeteries. In a slightly different account, it was Morris's idea that Abe and Miriam come to America, save some money, and perhaps return to Palestine with a financial stake. I do know that Miriam arrived first and got a job sewing in a sweatshop. Abe came a bit later, and when he arrived by boat, he was met by Miriam and Morris.

In one story, he was carrying a cane; in another, he had a branch from a tree in Palestine that he held behind his shoulders to straighten his posture. In any case, Morris—who was not a big fan of Abe's—took the cane or sapling, broke it in two, tossed it away, and told his future brother-in-law that Americans didn't carry such things around. And so, recounts my sister Laura, "Abe watched his one souvenir from Palestine float off into New York harbor." He also dropped the "sky" from his last name, and tossed that into the harbor as well. He was now Abe Rogow.

Most Jewish immigrants of Abe and Miriam's generation did not like to talk about the Old Country. They had fled poverty or persecution or both and wanted very much to be American, so they closed the door on their past. Many of my contemporaries know little or nothing about their ancestors, but I was lucky. Abe loved to tell stories about the old days and I loved to hear them, so I was left with an abiding interest in where and how he had spent his youth. In the fall of 1991, Cokie and I decided to visit a niece who lived in Vienna. Since the Iron Curtain had recently collapsed, we added Prague and Budapest to the trip. Once we did that, I insisted, we should go to Poland and see Bialystok as well. I didn't know how to begin, but I did remember vaguely that there was a Bialystoker synagogue on the Lower East Side of Manhattan, so I simply looked the name up in the New York phone book. I couldn't find the synagogue, but I did find an old-age home. So I called the number, and asked if anyone there could help me travel to Bialystok. The manager came on the line, said he was a native of the city, and gave me the name of an old friend, Anatol Leszczynski, who lived in Warsaw. Anatol was not Jewish himself, but he had studied the Jews of eastern Poland. I wrote to him and got no response, but when we arrived in Warsaw I had his phone number, so I called him up.

I'm an old man now, Anatol said, and when I got your letter I wasn't going to help you. But when I realized your family name was Rogowsky, I changed my mind. Alter and Shmuel Rogowsky were two of my father's best friends, they must be your uncles. So we engaged a car and driver and set out the next day for the three-hour drive to Bialystok. I wasn't sure what I was looking for. I just wanted some sign, some tangible evidence, beyond those faded memories and photographs, that my family had once lived in this area. A gravestone, perhaps, or a birth certificate. The first stop was Choroszcz, Anatol's hometown, only a few kilometers from Bialystok. On the main square was a house where Shmuel had operated a small candy shop. Another dwelling, down the road, was where Alter had run the Jewish burial society. We met an old woman who had known the Rogowskys. She told us that on Christian and Jewish holidays, her family would trade visits with Alter and Shmuel, and when times got hard for the Jews, they sometimes hid in her house. In fact, she added, she could still remember Alter singing Hebrew prayers on the Sabbath in a lovely voice. I was feeling pretty sentimental at that point until Cokie broke in: "That settles it, they couldn't be your uncles, nobody in your family can sing a note." She was right, they probably were not my uncles, but perhaps they were distant relatives. I'm still not certain.

Standing in the square at the end of our visit, I told Anatol that I had always heard that the family name came from a village in the area. (Abe sometimes told another version: Many Jews in the area only had first names, and when an edict was handed down, mandating them to adopt family names, some of them, as a joke, took on the name of a local dignitary, a Prince Rogowsky.) Much to my amazement, Anatol said he knew the place I was talking about, but he had to get directions. Soon we were headed down a dusty country road that couldn't have changed much in the eighty years since Abe and Miriam had left. Animals still powered many of the carts and plows. Remember, Rogow was my grandparents' name in America, and

the name on my birth certificate. Dad changed it to Roberts when I was two, but all my cousins still have the old version, and I've had those five letters imprinted on my brain my entire life. Suddenly, on the side of the road, we came upon a rusting metal sign with the name "Rogowo." There it was! My name! Not exactly in lights, but close enough. When I said I was looking for a sign that my family had once lived in the area, I didn't mean an actual sign, a roadside sign. But I had found what I was looking for. My roots.

We took pictures of me standing under the sign and then drove the five miles or so to Bialystok, which had been leveled by the Nazis during World War II and rebuilt as a drab, depressing place with little life or character. Even more depressing were the plaques that recorded brutal acts of anti-Semitism: Here was a field where five thousand Jews had been gunned down by the Nazis; there, the site of a synagogue burned to the ground with three thousand Jews inside. We made our way to the cemetery, largely untended and overgrown, but Anatol had made it his personal mission to keep clean a small plot surrounding a stone obelisk. On the monument were written the names of seventy-four Jews killed in the pogrom of 1906, the year before Abe left for Palestine. At that point I started to feel a connection to the place. My grandfather must have known some of the victims. Perhaps one or two were boyhood chums, joining Abe in that photo from the military school. The names on the monument were written in Hebrew or Yiddish, and when Anatol started reading them, I summoned up enough of my Hebrew school training to follow along. Fortunately, one was named Katz, an easy one to spot, where I could pick up the place, and as Anatol said each name, I pointed to it. I realized that, in effect, we were saying Kaddish, the Hebrew prayer for the dead. Abe's friends probably had no grandchildren to come back to honor their memory. They had died too young to have families of their own. So we were there instead, to pay our respects.

Then I asked Anatol if we could go to the railroad station. So many of Abe's stories had involved the station—picking up packages as a boy, leaving for Palestine, joining the army—that I wanted to see it. You're in luck, he replied, the station's one of the few buildings that survives from the old days, it probably doesn't look much different than it did when your grandfather lived here. When we arrived, I walked through the waiting room and onto the platform. Finally, I knew I was standing where he had stood. I could almost feel his presence. I had seen the city of his birth. I had seen the monuments to horror, to the relentless attempts to exterminate the Jews of this place. But Abe and Miriam had escaped, and built a new life in a new land. And almost exactly one hundred years after Abe's birth, his grandson had returned. I found myself saying, to no one in particular, "Pop, we survived, and I've come back to prove it." I returned to the car and, totally without warning, started sobbing. I couldn't regain my composure for twenty minutes. Something deep inside me had broken loose, something I didn't even know was there.

The trip left a deep impression on me. I wrote about it for *U.S. News*, where I was working at the time, and I was stunned by the reactions I received. Some writers wanted to share their own stories, others wanted help in researching their family history. My cousin Pam called, crying into the phone. We have so many things from my husband's family to show our kids, she said, and now we have something to show them about our family. A friend of mine, Frank Mankiewicz, offered this insight: For so many American Jews, the Holocaust is a wall that blocks us off from our own history. In the lands we came from, there's no one left to tend the graves or the records or the memories. I envy my Italian and Greek friends who can go back to their ancestral homes and even visit a cousin or two. I had no cousins left, but I was still very fortunate. I had managed to get past that wall, to reconnect with a past that had been stolen from me. A few years after our trip, the movie *Schindler's List* appeared, and as I watched scenes of Nazi soldiers, rounding up Jews for extermination, on the screen flashed the words "Cracow 1943." Cracow is a Polish city south of Warsaw, but the scene could easily have been set in Bialystok. And I was born in 1943. Now I understand, I said to myself. I was born in America to replace one of the lives lost in Poland at the same time.

On that same trip we also tried to find Eishishok, Miriam's hometown, but we had the wrong spelling and didn't realize it is now across the border, in Lithuania. Some years later, when Cokie and my sister Laura were visiting the Holocaust Museum in Washington, they had a shock. The central hall is adorned with hundreds of photographs of one small village, a village where two photographers had lived and worked and managed to save their negatives. It was Eishishok. Miriam had left a good twenty years before the earliest photos in the exhibit, but perhaps one of the photographers whose work was on display had taken the picture I still have, the one of Miriam and her family. With one exception, that picture is the only image that remains of any of my great-grandparents.

The only other picture of a great-grandparent shows a woman named Anna Meltzer, the mother of my mother's mother. I never knew her, but she and her husband, William, did immigrate to America in 1907, my only ancestors from that generation who ever saw these shores. Their daughter, my grandmother Sadie, was fifteen at the time. I know almost nothing about the Meltzers, except that they ran a dairy farm in Bayonne, and in the winter they delivered milk using a horse-drawn sleigh. They were observant Jews, living within walking distance of a synagogue, and they spoke little English. My mother has only one memory of her grandfather: sitting on his porch, as an old man, with a blanket over his knees. "I never thought of them as being part of my life," she recalls. Sadie Meltzer—I never knew her either, she died before I was born—married a man named Harry Schanbam, who delivered milk for her father.

I did know Harry; I lived in the same house with him until I went away to college. Like my other forebears, he was a political radical, a "flaming Socialist" in my uncle Bernie's description, who frequently ran afoul of the Cossacks, paramilitary units in czarist Russia, who specialized in beating up

Jews. To be specific, he was a "Bundist," a member of the Jewish Workers' Bund, a Social Democratic movement that was widely harassed by czarist forces in the late nineteenth century. Once, after he got kicked in the head by a horse, his mother dressed the wound with pumpernickel, a primitive form of antibiotic. And he was fervently antireligious. On Saturday, the Jewish Sabbath (or *shabbas* in Yiddish), Harry would walk the streets of his town, deliberately smoking cigarettes to signal his rebellion. And he would say provocative things like "talking to a Venetian blind" would do as much good as praying to God. I grew up with the story that Harry used to tell: "I finally left after the Cossacks rode through my mother's living room for the third time." How he came to Bayonne, or to deliver milk for William Meltzer, I do not know. Census records tell different stories: One dates his arrival in America to 1902, another to 1907. I do know that Harry and Sadie's third child, my mother, Dorothy, was born in Bayonne in 1919. Three years earlier, Abe and Miriam had had their first child, a son named Wilfred who was always called Will. My family was now in America, but the journey of becoming American was just beginning.

BECOMING AMERICANS

This business of becoming American has always been tied up with the issue of names. My grandfather Harry had two brothers in this country: Max, who ran a paint store in Jersey City, and Morris, a pants presser in the Bronx. The three brothers all spelled the family name differently, however. When my mother's brother Bernie, an orthodontist in Jersey City, considered changing Schanbam to something more American, his father always protested, "Who will carry on the family name?" What name, Pop, Bernie would reply, you can't even agree with your own brothers on how to spell it. (I've checked the census records, and in 1920 Harry's last name was spelled "Schanban" and in 1930 "Schanbaum." I'm not sure why, probably the census taker made a mistake, but confusion over names seemed to be a family trait.) Bernie and his wife, Lee, had one child, my cousin Seth, who died tragically of leukemia as a teenager. As a boy Seth came to his parents one day and said he wanted to change his name. "Thank God," replied his mother, whose maiden name was Davis, "I can't stand it!" She suggested possible alternatives, Sanborn perhaps, or Schenley. One the name of a popular coffee, the other a well-known whiskey. No, Mom, said Seth, he wanted to be Tommy Schanbam. "I was ready to kill the kid," recalls Aunt Lee.

Grandpa Abe had dozens of trades and schemes in a long and checkered professional life, but Grandpa Harry was different. After his youthful experiences delivering milk, he became a carpenter and a good one. I always marvel at that immigrant generation, leaving home and starting again, overcoming so many difficulties with such patience and persistence. What often kept them going was a sense of optimism about their adopted country. When Mom's uncle, Morris the Pants Presser, was close to ninety, he boasted to her, "Look at my suit, it's good material, it will last a long time." One of my earliest memories is of Harry's toolbox, made entirely of wood, almost two feet long, with a thick handle, stained dark with years of hard work. Just like him. Some years ago I found a box like Harry's at an antique shop. After I brought it home, it languished unused, lost in a large stack of baskets in a corner of our kitchen. But when I started this memoir, I cleaned it up and set it down next to my computer. Now it holds files and notebooks instead of hammers and nails, but it's still a toolbox.

When Mom was about five, in the mid-1920s, Harry built

the house on Thirty-first Street, less than a mile from where he had worked on William Meltzer's dairy farm. It was only the third or fourth house on the whole block, which still had a rural feel to it, and Mom remembers boys building fires on the shores of Newark Bay to roast potatoes. But new immigrants were flooding the city and rapidly turning the farms and waterfront estates into housing lots. By 1930, the population had swelled to almost ninety thousand and one-third were foreign born—the second highest percentage of any city of that size in the entire country. By the time my mother brought me home to Harry's house in 1943, the block was entirely built up, except for two lots at the end of the street. The dairy cows were gone and the whole city had a cramped, urban air.

Even though the Schanbams never changed their name, they adapted to their new country in other ways. Bernie became a star basketball player at Bayonne High, but his parents were horrified when they came to see him play. A son of theirs! Out in public in short pants! (Bernie was always a good athlete, playing golf until illness forced him to stop in his seventies. When he visited us in Bayonne, and I was out in the backyard shooting hoops, he'd astound us all by doffing his immaculate sport coat and swishing shot after shot from long range, never mussing his perfectly combed silver-gray hair in the process.) A favorite family story: When Bernie was in high school, he had a date, sitting in the stands for a basketball game, that he wanted to impress. So he asked the coach if he could be the captain for the evening, leading the team onto the court and catching the girl's eye. But during one scrimmage, he got hit in the groin and fell to the court in agony. The girl didn't understand what was happening, and out of embarrassment and delicacy, his friends told her he'd taken a blow to the head. After the game, when they all met up, the girl was very solicitous, asking about Bernie's injured head. He had no idea what she was talking about.

Sadie grew up in a religious household, but when she married Harry, she took on his rigidly secular ways. The Schanbams had no religious life or training of any kind. My mother insists that the first time she ever attended a seder, a ritual meal celebrating Passover, it was organized by her Catholic daughter-in-law, my wife, Cokie. The only matzoh—unleavened bread eaten during the Passover season—they had in the house was there for the Christian cleaning lady. Bernie was sent for a while to learn Yiddish at the Labor Lyceum, a center for Socialist-minded Jews, but he never learned Hebrew. This was a good metaphor for the Schanbams' approach to life in America. They were rejecting their faith but not their tribe. "They were not anti-Jewish, they were just antireligion, the same way I feel today," recalls Mom, and that mixture was not uncommon for Jews of that generation. They were Jewish the way others in Bayonne were Ukrainian or Polish or Italian. That was their identity, their heritage, their history. But they didn't always practice their religion, particularly if they were political leftists. If you asked my mother today, "What are you?" one of her first answers would be "I am a Jew." But she hasn't been inside a synagogue in many years. That's true as well for just about everyone else in my family, except for one person. Me. The guy who married the practicing Catholic. In fact, Cokie jokes, "You should have listened to your mother and married a Jew. She wouldn't have dragged you to temple or made you have seders."

Sometimes, of course, Jews wanted to deny or at least disguise their heritage. It happened to my own father. During his youth, particularly when he did business out of town, he would occasionally use the name Atlas instead of Rogow. Why Atlas I don't know. I suppose it sounded strong and didn't end in a vowel. When I was two, and Dad decided to change our last name permanently to Roberts, my mother was furious. He always said that Rogow was too hard to spell or pronounce, but she never bought his explanation. "He felt self-conscious" about his name, she says now. "He felt it might be a hindrance in terms of earning a living." It was 1945, he was involved in the New York publishing world dominated by "these Protestant characters," as Mom described them, and "he was trying to conform a little bit more." Mom went along, because it was important to Dad, but sixty years later she still resents the change: "I never wanted to feel guilty about denying my heritage. I was Jewish and I was pleased, frankly, that I was Jewish. Culturally, my heritage was as good as anybody could hope for." Frankly, I agree with her. I've always been proud to be Jewish, I never really liked going around with a name that is usually Welsh, and I'm sorry my father changed it. Like Mom, I've never wanted anyone to think I was "denying my heritage." In my heart I'm a Rogow, even a Rogowsky, not a Roberts.

But I had it easier than my parents or grandparents. Because of their efforts, my American identity was secure. I never felt self-conscious about my origins or was tempted to deny my heritage. Quite the opposite. I was comfortable enough to go back and reclaim some of the old ways. My parents called their four children Steven, Marc, Glenn, and Laura, all crisp, modern-sounding names (even if I was named for Grandma Sadie). My son is named Lee, for my uncle Bussy, and my daughter is Rebecca, a good Old Testament name. I once told my mother that if we had another son, which we never did, our favorite name was Samuel. She was horrified. It sounded like an aged uncle in the Bronx. But that's what I liked about it, and I'm thrilled the old names are now making a comeback. I recently had a student named Dorothy, my mother's name, the first person under sixty I've ever met with that name. Now, when someone calls out "Max" in the playground, at least three little boys come running. Enough with the Shannons and Heathers and Madisons. I'm old-fashioned and proud of it.

Here's the best story I know about Jews and names. I was once sitting at a bar in downtown Washington, waiting for a friend, and overheard two people chatting. They had obviously just met, at a conference or something, and the man was showing the woman pictures of his children. When she asked their names, he told her that he originally wanted to name his daughter Casey, but his mother had objected: "What kind of a name is Casey Klein for a Jewish girl?" The day the child was born, the man saw the name Courtney in the *New York Times* and decided that that was an acceptable substitute. After all, his mother had only objected to Casey. So what about your son, the woman at the bar asked. When he was born, the man answered, I called my mother and said proudly, "Your grandson is named Samuel Feldman Klein." And at that point his mother yelled: "What? You want everyone to know he's Jewish?"

In the twenties life was good for the Schanbams. They had their new house, construction projects kept Harry busy, the grocer from around the corner would deliver fresh rolls every morning. "You'd open the door, and there the bag would be," recalls Mom. And if you opened the door, you didn't see many strangers, either: "It was a small community, all of our friends were Jewish." One of them, Charlotte Blaustein, came from a wealthy family and a chauffeur sometimes drove her to school. The chauffeur's son was the only black in Mom's elementary school class. There was money for a car, a Studebaker with plastic curtains instead of glass windows, and occasional vacation trips. When Mom was nine or ten, the family traveled

My mother's parents, Harry and Sadie Schanbam, dressed for dinner at a Catskill resort in 1927. Harry had built the house on Thirty-first Street that I was raised in a few years before.

to Niagara Falls, staying at farmhouses along the way, since there were no motels yet lining the roadways, advertising HBO and hundred-item salad bars in garish neon. That trip carried great symbolic value. "Both of them were very eager to be Americanized, and they did a good job of it," Mom says of her parents. "This was like a journey into Americana. Americans took holidays." During other summers Sadie and the kids took a place in the Catskills, and Harry would drive up after work on Friday night, "Just like all the other American fathers." The only photo I have of Sadie and Harry together was taken in 1927, at a Catskill resort. They're dressed for dinner, he in a tie and jacket, she in a pretty dress. Just like all the other American couples. I keep asking my mother for stories about her family but she has very few. Her parents, like many others in the immigrant generation, just wouldn't talk about their past. "It's as if they wanted to close the door and put it behind them," she says today. "They were here now, and they would do their best to become Americanized and live like their neighbors."

Mom on the right, age eight, with her older sister, Sylvia. Taking vacations, says Mom, is what Americans did.

But their efforts didn't always pan out. Grandma Sadie's sister-in-law, Pauline, married to her brother Sam, was the family authority on all things American. So when Pauline arranged for Sadie to be inducted into the Order of the Eastern Star, a social and charitable organization associated with the Masons, it was an important sign of acceptance. For the ceremony Sadie put on a white dress, the required attire for the evening. However, the Schanbams lived on the second floor of a two-family house, and as she was walking down the steps, Sadie tripped and sprained her ankle. She never made it to the induction, and never went back to the Eastern Star.

Sadie's children use words like "basic" and "homespun" to describe her, so it's incongruous that in one of the few pictures of her that I remember, she was wearing a clown costume. The photo was taken at a Halloween party and probably Sadie made the outfit herself. Like the white dress for the Eastern Star, the clown costume was a statement of identity. Americans celebrated Halloween. Sadie was an expert seamstress, and Mom remembers the summer before her first or second year in high school, when Sadie worked for weeks making her a new dress, navy blue with a round neckline edged in white trim. Mom wore it on the first day of school, and recalls: "In honor of my new dress I went to school with lipstick for the first time." American girls wore lipstick.

Meanwhile, soon after Abe Rogow watched his walking stick, and part of his last name, float away in New York harbor, he and Miriam were married. Unlike the Schanbams, who lived a stable life in one place, the Rogows reflected Abe's restless and impatient personality. They left New York City for Newburgh, New York, where Abe found work on a farm. Dad, their first child, was born there on December 13, 1916, almost four years to the day after Miriam inscribed that photo to Abe saying, "You once had a true close friend." Dad always said he was originally named Sholom, after the great Jewish writer Sholom Aleichem, the author of the stories that became *Fiddler on the Roof,* who died the year Dad was born. Sholom also means peace in Hebrew, but it was too Old World, I guess, for an American baby. So Abe picked Wilfred out of a book because, he said, it meant giver of peace. Dad was always dubious about the story and used to kid that Abe's finger slipped when he was looking up the name. The writing gene in the Rogowsky code was still on Abe's mind when his second son, my uncle Bussy, was born a few years later and he was named Leon, for the Yiddish author Isaac Leib Peretz.

By this time World War I was on, and Abe heard there was plenty of work in Washington. He called himself a painter, even though he was never very good at it. But he was very good at studying a problem and playing the angles. Abe bought a motorcycle, and every morning he would show up at the union hall, where available jobs were posted. While the other workmen were finishing their coffee and figuring out where the job site was, Abe would hop on his motorcycle, beat them to the spot, and get hired. After a few days his deficiencies as a painter became obvious and he was often fired, but he earned a few dollars and gained some experience and went on to the next job. The motorcycle came with a sidecar, but when child number three, Rose, arrived, things got crowded. Abe rigged an extra seat, a bit like a hammock, where Dad rode proudly while Miriam, Rose, and Bussy crammed into the sidecar. Like many of Abe's ideas, it was brilliant but flawed. One day, when Abe took a corner too fast, Dad fell out of his seat and cracked his skull. Then, around 1920, the economy dipped and work became scarce. Abe bought a set of tools and proclaimed himself to be a carpenter, a "Sears and Roebuck carpenter," in the adage of the day, but he was having trouble making ends meet.

Miriam had a cousin in Bayonne named Joe Wolansky who was doing pretty well. At the south end of the city, right across from Staten Island, there was an amusement area called First Street, a sort of minor-league Coney Island with rides and refreshment stands and penny-ante gambling games. Some games involved skill, throwing a ball or tossing a ring. Others were pure chance. A noisy, colorful wheel would spin around and land on a number. If you bet on that number you won. Since the prizes were mostly junk of the "kewpie doll" variety, the profit margin was decent. Joe ran one of those games and invited Abe and Miriam to join him. So they packed up the kids, took the train to Bayonne, and started yet another new life, this time in the amusement park business.

Wolansky divided his stand in half with a blanket, and Abe rented out one side and set up shop. His view was that people would place more bets if the prizes were upgraded, so he went looking for better merchandise. Two of his finds were hand-wound record players and pocket watches. The traffic increased, and Rogow and Wolansky branched out. They were basically shysters. They bought sugar pills and sold them out of two vats as "pep" pills, one labeled "regular" and one "extra strength." Both were the same, both useless. They actually hawked a kind of snake oil, which promised to cure all sorts of ailments. And then there were the freak shows. Wolansky's nephew Sam Sorkowitz had dropped out of school and run away from home to join a traveling circus. Now he was back at First Street, operating a tawdry little bunco game, The Head Without a Body. It was all done with mirrors. No kidding. Someone sat on a stool and stuck his or her head through a round hole in a triangular table. The sides were draped in black cloth and mirrors were arranged in such a way that it looked like the head was floating in space. Various people took turns playing The Bodyless Head, including Sam's brother Julius, who recalls: "I did it quite often, but you couldn't move around, you were stuck in there." One enterprising girl introduced a new wrinkle. She became The Singing Bodyless Head, and the patrons, who had paid an entrance fee, would toss extra coins her way. Julius was helping out one day when one of the coins clanked against a mirror. "We thought there'd be a riot," he recalls. "Sam whispered, 'Get ready to run,' but nobody caught on." They didn't catch on, I suppose, because they didn't want to. Where was the fun in that? But Julius has another take. The clientele were mainly unlettered immigrants who worked menial jobs in Bayonne's growing industrial plants. "You'd be surprised how unsophisticated people were in those days," he recalls. "They'd fall for anything."

Near the old First Street complex where Abe got his start was an undeveloped tract of land owned by a man named Hermaneau. He had leased the land, intending to build a new amusement park, but he knew little about construction or promotion. Abe didn't know much either, but he had plenty of drive and energy. The two men formed a partnership and built the Bayonne Pleasure Park. For once, Abe's timing was right. The Roaring Twenties brought a new wave of prosperity. Operators poured in, paying rent to open up stands, rides, and snack bars. The partners built a large hall on a pier, called the Bayonne Casino, and there were attractions every night—basketball games, roller-skating contests, dances, even prizefights. Julius remembers selling Eskimo pies between rounds of the fights when he was only eight or nine: "The fights were a big thing and the park was a very busy place, they came from all over New Jersey and Pennsylvania."

The Rogows were suddenly rich. At the end of the season, Abe and Miriam would go to Havana and gamble. They bought a lot on the fanciest street in Bayonne, Wesley Court, and built a new house featuring all sorts of Abe Brainstorms. One, miniature bathroom fixtures for kids, proved workable. Another, unstained cedar shingles, didn't. They were supposed to turn silvery in the sun, but it never happened. Abe always insisted that he had introduced another brilliant innovation on Wesley Court, built-in kitchen cabinets. I find it hard to believe that Abe was the first to come up with this idea, but in his mind just about everything he did was The First and The Best. A fourth child was born as the house was being built and he was named Morris, after Miriam's brother, who had brought her to America and then died in the flu epidemic of 1918. We always knew him as Morrie or Murray.

The house was ready in 1927, the high point of family fortunes. It featured a pool table, which drew kids from all over the neighborhood. There were always houseguests, Abe's old buddies from Palestine or distant relatives, newly arrived in America and needing a place to stay. Often, they worked at First Street as well. That's what families did, they took care of their own. However, two years later the stock market crashed, and so did the Rogows. Nobody had extra money for merry-go-rounds and freak shows and prizefights. The Bayonne Pleasure Park fell into disrepair, the fight arena burned down, and since Grandpa had no insurance, it was a total loss. (He later learned the virtues of underwriting and was known to say of a friend, "He deserves a good fire." He meant, of course, that sometimes a well-timed insurance payment could exceed the worth of a building.) Abe kept a few gambling games open, but it wasn't enough. So Miriam set up a stand in First Street selling clothes, mainly for women. She had done this work as a girl in Bialystok, but now the family was desperate and she was relentless. "She could con the customers," Mom remembers with some admiration. "The dress might be wrong, but she could talk anyone into it." Adds Aunt Rose: "She was out there punching, she had tremendous energy."

But standing on her feet from four in the afternoon to two

in the morning, selling her heart out, took its toll. Miriam was physically exhausted and often slept till noon. The house and the kids took care of themselves. Murray remembered going to school an "absolute mess" and being sent home. At age ten he was taking the bus to New York by himself, seeing stage shows, and "eating five Baby Ruths" for dinner. "I was allowed to run free," he recalled. The one household chore Miriam always liked was cooking, so there was plenty of food at every meal. Murray said of his mother: "She'd get upset unless everybody stuffed themselves to the eyeballs." My uncle battled extra weight his whole life and once, after a bit too much psychotherapy, stormed into the family house and confronted his mother. I wanted love, he cried, and you gave me food. You never let me grow up. Miriam shouted back with tears in her eyes: "So grow up! So who's stopping you?" I was with Miriam on that one.

Miriam also helped support the family during hard times with her card playing. She called her opponents "pigeons," and in later years she would often convene a game at our house when she was taking care of us. This was not a ladylike round of bridge or canasta. It was poker, and it was for money. Waking up and walking to the bathroom during the night, I had to pass the doorway to the kitchen, where my grandmother presided over a spectral scene. The kitchen table was covered in a glistening white cloth. The light above the table was turned to its brightest setting and pulled down low on its movable chain. Coils of smoke filled the air as the players puffed away on filterless cigarettes. Looking through that doorway and into that room was like peeking into another world, intriguing and yet unnerving. How could our kitchen, so familiar during the day, look so different at night? But we liked Grandma's visits. Since she always played with new cards, she left the used decks behind for us. And since she was too tired to clean up after a long night of intense competition, snacks and drinks were often left over as well, bowls of peanuts and half-consumed bottles of Canada Dry ginger ale.

So Dad grew up in two worlds. One was pretty sleazy, a world of con games and card games, barkers and braggarts, gambling and hustling. He never liked it much, but as the Depression deepened, he'd get roped into working at the First Street stands. It was the only way he, or anybody else in the family, could make any money. Once he got into a fight and came out very much on the short end, with a separated shoulder. But the injury later kept him out of military service. And his First Street years left him with one great talent: He knew how every game in every carnival was rigged. If you went to a county fair with Dad, you were sure to come home with the biggest stuffed animal in the place. And he could guess anybody's weight within five pounds. As a kid, this was better than having a dad who could hit a baseball or throw a pass!

Dad's other world was an intellectual one. After all, he was named for a famous Yiddish writer. His father was largely self-educated but well versed and highly opinionated when it came to politics. Young Willie was very smart in school and got pushed ahead rapidly, much too rapidly. He graduated from high school well before his sixteenth birthday and enrolled in New York University as a commuter student. He might have been ready for the schoolwork but not for the social life, and his mother didn't help matters. Using her connections in the clothing business, she bought him a box of six shirts to outfit him for college. They were wholesale, the price was certainly right, and money was tight. But they were all the same color. Day after day he'd take the bus into New York from Bayonne, wearing what appeared to be the same shirt. Soon his friends started talking. Willie, what's with the shirt? It was not easy to explain that in fact he had six shirts, not one. They all just looked the same.

Like his father, Dad had a penchant for left-wing politics. He sometimes described himself as a "Lovestonite," a follower of Jay Lovestone (born Jacob Liebstein), an early Communist leader who broke with party leaders and advanced the view that American capitalism was different from Europe's and not subject to the same rules of collapse and decay. NYU in the thirties was, to say the least, a turbulent place politically. Like their ancestors back in Europe a generation earlier, many children of immigrants—faced with no jobs and little hope—turned to Marxism as the answer. There were as many splinter groups as there were coffee shops in Greenwich Village, and Dad used to tell the story of one small faction that had been infiltrated by two Communist spies. When Dad warned the leader of the group about the traitors in his midst, the guy replied: "I know, I know, but they're the only ones who do any work!" He was also very antiwar, and helped write an American version of the famous Oxford pledge, taken by British students in 1936, who vowed never to fight "for king or country." Like Abe, Dad was seldom quick to admit a mistake, but on this one he would always say, "Thank God no one listened to us." Even after the war, however, Dad clung to his left-wing inclinations, voting for Henry Wallace in 1948 and staunchly believing in the innocence of Alger Hiss.

By 1936, the Depression was hitting Bayonne, and my family, with full force. Mom entered NYU, but the Schanbams' finances were increasingly shaky. The Rogows lost the fancy house on Wesley Court and had to move into cramped quarters only a block away from the Schanbams, on Thirty-second Street. During the good times, the Rogows had employed a full-time maid, a Polish woman named Sophie who spoke little English. When they left Wesley Court, they didn't have the heart to tell her that she no longer had a job. But Sophie tracked them down and showed up at their new house one day. Finally, the Rogows had to admit the embarrassing truth, they could no longer afford her services. Dad was still in college, but increasingly confused and frustrated about job prospects. A philosophy major, he tried law school for a while and gave it up as a bore. As a last resort, he focused on getting a teaching degree. The future was looking pretty bleak. But in February 1936, Dotty Schanbam's friends decided to give her a seventeenth birthday party. They called a fellow named Eddie Adler (I later went to grade school with his daughter and nieces) and asked him to bring some guys to the party. One of them was Willie Rogow.

LOVE IN THE RUINS

Will was the life of Dorothy's birthday party. He told a long and amusing tale about going to Philadelphia that ended with the punch line "And that's where I met my future wife." Almost seventy years later, Mom tells the story with a happy smile: "Well of course, in retrospect, it was a very strange coincidence." At the end of the evening, Will offered to walk the birthday girl home. But she soon discovered that this jolly and talkative fellow turned painfully tongue-tied when he was alone with a woman. Even though it was a cold and icy night, he abandoned her three doors from her house and bolted away. "He was obviously very smart and very funny and very shy," Mom remembers. "A lot of contradictions." True, he was always a man of contradictions. From that night onward, however, he never wavered in his devotion to Dotty Schanbam. "I think I was the only girl he ever took out," says Mom, and I'm sure she's right. Dad used to love to gaze around at family gatherings, surrounded by a clan that reached eighteen people before his death, and say, "See what happens when you walk a girl home from a birthday party. . . ."

Dad at about age twenty-five, a year or two before we were born. He grew a beard while recuperating from an auto accident that broke his pelvis.

Mom and Dad lived a block away from each other, they both rode the bus into New York for college, but they didn't see each other for almost a year. On New Year's Eve, as 1936 turned into 1937, a friend of hers from NYU held a party and told Dotty "to bring a fella." So, she recalls, "I got up my courage and called him." At my urging, Mom unearthed a box she'd squirreled away in the basement. It contained dozens of letters she and Dad exchanged before they were married. When I started reading them, I was stunned to see the earliest postmark: April 1936, just a few months after they'd met, and long before the New Year's Eve party. She'd forgotten about the letters, she said, but as we talked she remembered. "It was much easier to express ourselves honestly by writing. It was just a very easy way to get to know each other. I think it was a great idea."

She was right. The letters that flowed between 174 W. 31 St. and 129 W. 32 St. reveal two shy and sensitive souls, sophisticated about books and innocent about life, slowly getting to trust each other with their most hidden thoughts and tender feelings. They're like two birds, flying around in circles. Once they landed on the same wire, they were never apart. I'm not sure who wrote first. The earliest letter that survives is written by Dad, but he makes reference to a previous note. It is postmarked April 7, 1936, at 7:30 P.M., and he's so unsure of himself that he got her address wrong. He originally wrote "34th Street" then crossed it out. Apparently, he also put the wrong stamp on because the envelope is marked "Postage due 2 cents." His tangible timidity continues with the salutation: "Dear Dorothy (or Dotty, or Dot, or whichever of the variations of your peculiarly flexible name you prefer)." He mentions that the Rogows have just been forced to leave Wesley Court, and in the move "I find I have misplaced two pipes, Kant's Critique of Pure Reason, a penknife, the June 1932 issue of 'Paris Nights,' an ax handle, and the letter I wrote you in regard to your critical and poetic efforts." Apparently, she had called him "gallant," an overly generous use of the word, since he had left her three doors from home, and now he was expressing his "indignation." The term "gallant," he blusters, "implies a whole false standard of hypocritical chivalry, division of sexes, pedestals, and martyred masculine servilance." There's no such word as "servilance," he probably meant "subservience," but in any case the letter goes on for ten handwritten pages in a similar vein. I'm not sure whether he's trying to be funny, or just comes off as stuffy, and he isn't sure either. He admits on the last page that if he reads over the letter, he'll be too nervous to send it. So he simply concludes: "What the Hell. Falteringly, Will Rogow." Oh, Dad, what a dashing romantic you were!

She writes back six days later, and it's clear, even then, that she understood him very well. At seventeen, she was already a woman of strong opinions who did not hesitate to express them. "I wonder," she writes, "if there's a psychological explanation for your resentment of the female of the species. Or, are you such a perverse individual with deliberate intentions of approving nothing the multitude does?" Then she answers her own question: "Your whole attitude reminds me of the little boy whistling in the dark to hide his fear." Bingo! She saw right through Willie Rogow, the boy who went to college at fifteen wearing the same shirt every day, who understood everything about Kant and nothing about kissing, who read voraciously about nights in Paris, but left a girl standing on a treacherous street in Bayonne because he was too nervous to walk her to the door. His reply, started at 2:00 A.M. and finished at 4:20, recalls his reaction to getting her letter: "About fourteen hours ago, my complacency and self-possession completely crumpled," and only now am I "beginning to recover my composure." He goes on for another ten pages, discoursing on chivalry and gender equality, but on the last page he gets to the point. "When you described me as a 'whistling, frightened boy,' you happen to be quite right. . . . I am exceptionally reticent and bashful in connection with women. I have a pretty glib manner, but that's as far as it goes." A man of contradictions, as Mom described him, and he reinforces that point with his own words: "Thus, while I hold extreme and advanced views on sex-morality, views which from a Puritanical slant are highly immoral, I am the most completely chaste and virginal fellow you ever came across." And he concludes: "Yes, Dot, I'm very often, whistling in the dark."

In her reply Mom reveals a lot about herself. She's clear-minded about who she is and what she wants. Her ambitions are more limited than her abilities. "As a result of my admittedly few years of observation, it's my opinion that, fundamentally, what most women want is only an appreciation of their mental capabilities, in the abstract," she writes. "Most women are not very desirous of taking up the whip and taming the world. What they want is to make men aware of their potential ability to do so, though. (This must seem like treason in the ranks!) To express this idea in another way—they want to be sure of equal opportunity without the faintest notion of taking advantage of it."

They weren't married for another four years, but it's obvious from those letters that something was already starting to grow, and glow, between them, a small spark of candor and intimacy. She sensed his deepest fears and brought them into the light, where they weren't so scary anymore. He found her insights painful but perceptive, and he knew he needed them. She wanted him to take her seriously, but she also wanted him to know that she'd never threaten or compete with him. Once she became pregnant with Marc and me, she never worked a day outside her home. The outlines of their lifelong partnership were already coming clear. He was the glib and amusing one, the wind in the sails. She was the strong and steady one, with her hand firmly on the tiller. He was the outsider, their face to the world; she was the insider, the beating heart of the family. It was that way for the next sixty-one years.

Young love was blooming, but the Depression was choking off their sun and water. If they wanted to go to the movies, they would "scrounge around" for used bottles they could return to the store for pennies and nickels. Most of the time, groups of young people would simply gather at each other's homes. They had one rich friend, whose family sold building supplies, and since he owned a fancy record player, his house was a favorite destination. One night they went to a local movie theater and Dad was thrilled to win a contest, a raffle of some sort. The prize was a phonograph he couldn't afford to buy on his own.

Dad responded to the economic gloom by spouting Marxist rhetoric. "Viewed in a historical sense," he wrote in one letter, "we are living through a stage of social disintegration and decay. Old standards of value, traditional morality, all are breaking down and proving inadequate. There is a lessening of opportunity, a dying of hope, and resultant feelings of futility, fatalism, confusion, etc." Mom felt this futility and fatalism in a more direct and personal way. She says in reply that "it's a bit defeating to reach for the moon" and realize that you have "only limited opportunity to pave your way."

She concludes with the faint hope that "I might find a niche—better to say, hole—for myself as a teacher."

This sets off a furious exchange. He accuses her of retaining "petit-bourgeois standards of value." She allows "the white heat of my indignation" to cool down before responding: "Believe me, Will, the prospect of social and economic prestige is *not* my guiding star." But, she adds, "a certain monetary return would, I admit, be essential, because I have a certain amount of pride that revolts against my being dependent on my folks for financial support any longer than is reasonable." He backs off slightly, writing about his career prospects: "I feel rather futile on this particular score myself." But then the Marxist/Bohemian/Utopian theology kicks back in. Making money, he says, "stands so low in my standard of values, that I don't feel very fatalistic as a whole." She tries to smooth things over as well, writing on NYU stationery, "I've been thinking that perhaps I didn't take your criticism as a 'good sport.'" She apologizes for the short letter, pleads cramming for six exams as an excuse, and ends, "I want you to know, though, that I'd like this correspondence to continue. For one thing, it's doing a lot toward developing my critical intelligence. . . ." Is she tweaking him a bit? Of course. I wonder if he got it.

Earning a living might have been low on Dad's priority list, but he had no choice. He needed money and there was only one place to earn it, First Street. He writes to Mom in midsummer of 1936, after a month of silence: "I have been busy from the moment I've opened my eyes in the morning (one o'clock) till I close them at night (anywhere up to four ayem)." This includes, he adds, a five-hour stint "actually operating the combination revival meeting, medicine show, auction house and food market that is my stand." He admits to getting "a certain kick out of the game," but in the end, the time on First Street has been "one of the emptiest barren months I have ever spent."

Back at school in the fall he found work he really could love, starting a magazine, *The Square*. He later called it "the most important experience I had at NYU" and added: "This helped develop an interest in writing, editing and publishing, which later became my career." He can't help showing off, and writes to Mom on the magazine's stationery, which lists his name as editor: "While I can see an objection to this stationery on aesthetic grounds (it really is quite ugly) you'll pardon the seeming exhibitionism which prompts its use. I'm an editor with an office, a telephone (in a few days), an editorial staff, a semi-secretary, and a hundred and sixty dollars to spend, unanimously granted by the Student Council. Maybe someday I'll even have an issue being sold."

I don't know how many issues were published or sold (the cover price was ten cents), but I have the first one, dated Winter 1937. *The Square* refers to the center of NYU's campus, Washington Square, but it could also describe the social status of the editor. In a commentary he contributes to that first issue, Dad writes that the "collegiate myth" spawned by the "jazz age" of the twenties featured "drunken youths driving in high-power cars, co-eds with revealing skirts, and breasts popping from low-cut dresses, immorality, promiscuity, raccoon coats and an utter lack of any intellectual qualities or sober thoughts." The Depression had changed all that, and while he praises the modern student for being more "sober, industrious, intellectual and literary" than his predecessors, he clearly misses the high-power cars and low-cut dresses of an earlier era. Just one breast popping briefly out of one garment would have been plenty! In describing the typical NYU student of 1937, he's really describing himself: "He lives at home. . . . He knows nothing of the reputed immorality of college dormitories. None of the mysterious orgies of fraternity are his to revel in." Instead his life is defined by rush-hour crowds on the subway, antiwar demonstrations, "and of course the Cafeteria." On some days, when money was particularly tight, Dad's lunch in that eatery was a "ketchup sandwich," a thin film of red tomato paste between two slices of bread. And for all his fantasies about "co-eds with revealing skirts," his heart was already promised—though he didn't quite know it yet—to the girl from the next block, Dotty Schanbam. Mom remembers visiting him one day at *The Square*'s offices, but she was too shy to go alone. "I had my bodyguards, two of my girlfriends came with me," she says.

By the spring of 1937, Mom and Dad were seeing each other regularly. But their financial insecurity was matched by their social insecurity. A few years later, Dad recalled in a letter their first date: "I thought of the first time I took you to the movies, to see 'A Nous La Liberté.' I was scared, honey, but felt that you would make things easy for me. I remember imagining all sorts of situations in which I would not know what to do. I remember worrying all day as to what I should do if you insisted on paying your own carfare. Honest! I don't know where I got that idea from, but it occurred to me, and had me stumped." He remembered holding her hand, of feeling "quite proud of myself" for settling her coat on the back of her chair. "I think what really frightened me most was what would happen after we left the theater. I was a mighty relieved young man when we got on the bus to Bayonne."

Their letters dwindle, but they still find it easier to argue on paper than in person. He keeps accusing her of a bourgeois tendency to overvalue social and economic status, and she finally rips back: "I can't quite understand how I left myself so wide open for misunderstanding. This has a familiar ring—I'm convinced I've said it before—to you. Two things that *don't* trouble me are the economic and social." But amid all the familiar exchanges about ideas and ideology, a new word creeps into their letters. Love. Mom confesses to being antisocial ("Generally, I don't like people!"), but then adds in the same letter: "The one thing that still remains with me, contradictory though it may seem, is the hope that someday I'll find love. . . . Childish and idealistic? Well, maybe—but it's important to me. That's my happy ending."

That's his happy ending, too, of course, but he's still scared to death, so when she first brings up the word, he deflects it with a literary reference. "You do mention one value, love," he writes. "I have always regarded love as one of the major values, especially after reading D. H. Lawrence." Until that point, what he knew of love came from reading, not reality. He'd probably never met a coed in a revealing skirt or a low-cut dress—let alone gone out with her—and the safety of a letter allows him to confess his feelings of inadequacy. He bemoans his inability "to master the fine points of human relationships, especially relationships with women." The frightened boy, still whistling in the dark, concludes: "You see, Dot, I wouldn't have the nerve to seduce you if I wanted to. Maybe that's what I've meant all along. Hell—I don't know. Will."

At some point that spring, about fifteen months after their first meeting, they share a kiss. But it didn't go well. Dotty writes: "There's a matter that we should thrash out. Your parting remark last night provoked me considerably." Apparently he had said something really suave ("Thank goodness you don't expect any pretty speeches") and she was furious: "You were looking forward to a cataclysm and I to the natural process of things. And although the natural thing happened, you seem to regard it as a catastrophe." Don't worry, she writes, "it's not likely to reoccur." But then she softens: "How my guardian angel must be laughing and saying to herself—'What a fuss over a few kisses.'"

Their meetings, and their kisses, did recur, but soon he got cold feet, telling her it wasn't such a good idea to see each other so often. She dismisses his fears as "fallacious, in view of the fact that I know of few other ways I'd prefer spending my time—that I know few other people I'd prefer being with." She then recalls that he keeps asking her what she thinks of him, so she gives this answer: "Is it enough to say your utter fineness is something I'll always remember, that your complete honesty in what you think and do has made me an entirely different individual? If not—kiss me, you punk, and find your own answers."

It's clear who kept the relationship going. She met his fear with fire. But it wasn't easy, living at home and being in love and trying to find a space and a place of their own. For a time they blacked out the return addresses on their envelopes—I can't imagine who they thought they were fooling. And they developed a code word to express affection over the telephone in case other people were listening, a pretty good bet, given the crowded conditions they both endured. Here Dad's pedantic side was at war with his romantic side. Their code word was Schopenhauer, the name of a nineteenth-century German philosopher. Perhaps Dad was being playful after all. Schopenhauer's most famous work was called *The World As Will and Idea.*

As their affection grew, their debates continued. He argues for the importance of sexual experience before marriage. She clings to notions of "old-fashioned" morality. But she wants to keep seeing him. As the school year ends she writes coyly, "Would like to return your workbook to you. Drop around some afternoon at home." When he graduates a few weeks later, she writes again: "I have a vague idea of what you want of life and living. Whatever it is, Will, I hope it will someday be yours." He thanks her for the note, and reflects on other well-wishers who have complimented him on his achievement: "Sometimes, I actually listen to what they say, and when I realize what they wish for me, 'success' in the best middle-class tradition, I think 'god forbid.'" He signs the letter, "Will B.A." and adds: "No, not showing off, just being a wise-guy."

A poor wise guy. For all of his protests against "middle-class tradition," he needed a job and couldn't find one. He did have a teaching certificate but no political pull, and pull was necessary. Two of Mom's friends were also looking for teaching jobs at about the same time, a pair of cousins, Sarah Hammer and Anne Botwinick. They were the nieces of William Rosenthal, one of Bayonne's richest men, the founder of the Maiden Form Brassiere Company, but with all his influence, he could swing only one job. It was decided that Anne would get it because she was less likely to snare a husband and needed it more. (Anne did get married, and had a son, and by the time she taught me high school history, she was one of the finest teachers in the city.)

With no other options, Dad was drawn back into the world of carnivals and gambling. "Sometimes we made a little money," he wrote years later, "but it was all a dead end." He and his brother Bussy spent the summer of 1937 in Wildwood, a seaside resort in New Jersey, running a concession of some sort. He makes friends with a pair of astrologers who run the next booth over and they take a day trip to Atlantic City. But he hates the work and writes on a postcard: "You could not distinguish me from the driftwood which wanders about this seaside paradise." He finally sends her his address and he's using the name "Atlas." She writes back that her summer is just as dull: "Did you ever realize how wonderful Bayonne is—as a place to die?" By August, he finally admits: "Dotty, I miss you. I am realizing increasingly how I enjoyed being with you, talking to you, listening to you."

Then the summer was over and the resort closed down. There were still no jobs in Bayonne, and Dad spent a lot of time hanging around with his friends in a local park. As the Rogow fortunes continued to decline, so did the Schanbams'. Mom's brother Bernie was headed for dental school and had first call on the family's shrinking resources. Harry built three houses in Jersey City and told Mom that if one of them sold, he could afford to keep her in college. Dad sends her a hand-delivered note in November addressed "Via Dog-Sled and Special Courier." On the envelope he draws three crude houses with a sign on one saying, FOR SALE SEE H. SCHANBAM. He concedes at the bottom that the silly drawing represents a "reversion to adolescence," but his cheerleading did no good. None of the houses sold and Mom had to drop out. Her friends helped get her a job at Maiden Form, in the record-keeping office. It was mind-numbing work, but she earned $21 a week, the only regular income either one of them had. In the spring of 1938, her mother, Sadie, died from cancer. Dorothy was nineteen, and the future looked gloomier than ever.

That summer Dad took a brief trip with his family to the Jersey shore. By then he's calling her "sweetheart" and she's calling him "darling." In a note he addresses a theme that recurs off and on for many years: his desire to write. With no teaching jobs available, and no interest in the gambling life, he reaches for something that runs deep in his background and breeding: "I am going to try to write tomorrow. I've been mulling over a story or two, and I think I can get them down, at least in a first draft." His subject: "I'll try to catch the self-conscious struggle for a good time and a vacation among the rooming house denizens, the attempt to concentrate as much excitement and glamour into two weeks as the surroundings will allow—and the resulting commercialized entertainment, people making money out of people trying to buy fun." Reading those words makes me sad. Dad had the writing gene but not the sitting gene. He had a good eye and a lively imagination, but lacked the self-discipline to translate the thoughts in his head to words on paper.

Mom always knew this about him, and kept chiding him for not actually writing the stories he kept talking about. Later that summer he was running a bingo game for a friend of his father's at an amusement park in Connecticut, outside of New Haven, the site of Yale University. "I think you'll get your story soon," he writes. "My room is rather quiet, and there are a few ideas buzzing around. You must be patient, however, as things are rather vague." One day he walked the six miles into town, saw the university, and huffed that the buildings reminded him of "night-marish enlargements of St. Vincent's Church in Bayonne." The old leftist leanings come spurting out as he derides the Latin inscriptions and the names of the rich donors adorning the buildings. "The whole thing seems to be a monument to endowment," he writes. But the commuter student who never went to a frat party or owned a raccoon coat can't quite cover his sense of envy.

With the summer over, Dad was back at First Street, working for his father again. Writing one evening, before business picks up, he promises: "Dotty, honey, some of the first stories I shall write will have the denizens of this 'international amusement resort' as the characters." He goes on to describe some of them: The customers who come "night after night for thrill, excitement, noise, people, and the hope of winning a lot of money"; the hangers-on, "whose whole life has become sitting on the benches and exchanging meaningless political gossip and boastful memories." Then he boasts a bit himself: "I'll write about them all some day."

But he didn't. And even the trickle of income produced by First Street was in danger of drying up. Political tensions in Bayonne threatened to close down the park, which always survived on the edge of legality. Mom and Dad wanted to get married, but they were each still living at home and couldn't afford a place of their own. So, like Abe and Miriam, who planned on meeting in Palestine and starting a life together, Will and Dorothy came up with a similar scheme. Except their Promised Land was in a different desert, in the Old West, not the Middle East. Gambling was legal in Nevada, and while Las Vegas hadn't been developed yet, Reno was a thriving tourist town. Abe had an idea for a new game, a form of roulette wheel that paid cash if you bet on the right color. Convinced, in typical fashion, that his invention couldn't miss, Abe put up $500 to finance the venture. When Dad hit it big, he'd send for Mom, and they'd get married and live in Nevada. So, in late March 1939, Dad got on a bus and headed west.

OUT WEST, BACK EAST

When Dad left for Reno, he was twenty-two and Mom was twenty. They wrote close to one hundred letters to each other over the next four months, and I have most of them, a poignant record of young love and hard times, high hopes and dashed dreams. They never talked much about that period, I guess because it was just too painful, and the only souvenir Dad brought home that I remember around the house was a metal neckerchief slide in the shape of a steer's head. But Mom kept their letters all these years, stacked neatly in chronological order, so I guess she wanted someone to read them. And when I did, I learned a lot.

The first thing was their passionate devotion to each other. Their relationship had started slowly and developed gradually, from acquaintances and neighbors to friends and

Mom and Dad at the country home of a relative in Mount Freedom, New Jersey, in about 1945, when his book business was just taking off.

then soul mates. The young man who once whistled in the dark to hide his fears about women was now more ardent than anxious, at least with Dotty. As he wrote in recounting their first date, she had made it easy for him. She had given him a gift, of confidence and reassurance, which he unwrapped eagerly. He, in turn, had shown her that her "childish and idealistic" dream of finding love was—literally—right around the corner. They were not formally engaged, but they were "spoken for," in Mom's wonderfully quaint phrase, and they would occasionally sign their letters "Your Husband" and "Your Wife." Here's the beginning of a typical letter, written several weeks after he arrives in Reno: "Dotty darling, I opened my eyes this morning and saw the empty pillow beside me. I closed my eyes and tried to see your dear face close to me, so close that I might kiss you, gently so as not to wake you. Just thinking about what is to be warmed my heart and sent me falling desperately in love with you all over again."

From the letters, it's also clear how suffocated they felt in Bayonne, and how heavily the Depression weighed on them. Mom hated the fact that their hopes for a life together rested on "man's weakness for gambling," and many years later she was still offended: "This was sleazy stuff. Who needed it? Who wanted it?" In one letter he notes that the father of a friend—the one who had introduced them—had been arrested for bootlegging, and that Abe himself had served a few days in jail. But this "sleazy stuff" was their only way out.

Dad writes about "the black void of the future," of the "desolate misery" they felt before he left. "I remember when you were dejected at the emptiness ahead of us," he recalls, in reminding her why they took such a chance on Reno. And they describe how hard it was to be living at home, in such tight quarters that Dad could tell each of his siblings by their distinctive nighttime noises. "I can see you looking at your kiss, smeared on my face and telling me to wipe it off as someone is coming," he writes. And he reminds her of the time that they stole an embrace in his parents' house and his younger brother, Murray, made her blush by noticing "the tell-tale marks on my face" during dinner. "Darling," he assures her, "soon we shall know the thrill of looking at a room and realizing that it is ours, the four walls enclosing a bit of space for the exclusive use of Mr. and Mrs. Rogow. . . . If anyone comes they'll have to ring a bell, knock on the door and set off a fire alarm, but we don't have to answer unless we want to." They weren't asking for much, but even a room of their own was beyond their reach. Unless Reno worked out.

As Dad heads west, his letters are full of hope and optimism. Writing on the stationery of the Plains Hotel in Cheyenne, Wyoming—adorned by a photo of "Chief Little Shield" in a feathered headdress—he talks about meeting travelers who assure him that you can start your own ranch for an investment of "a few hundred dollars." He's obviously misinformed, but he wants to believe their words so badly: "I think that we're going to own a ranch, darling. It's easier than I thought." The landscape is so new and unsettling that he searches out familiar landmarks, like a Jewish name on a clothing store in Laramie. "I dropped in [and] sure enough, a real *landsman*," he reports, using the Yiddish phrase for a fellow Jew. "I don't think there has been a single town of any size in which I haven't seen a sign 'Ginsburg' or 'Weinstein' or some such name, almost always on a clothing store and pawn shop."

Arriving in Reno, he checks into a place called the Hotel Golden, using the name Atlas, "a convenient handle for business dealings," and assures her: "Many people have similar dual names for business, often to cloak a Jewish name." He begins looking for space to set up his gambling wheel and meets a man named Smith, who runs Harold's casino and seems interested in Dad's proposition. But a day or two later Smith says no, the game would cause too much noise and commotion. "It looks bad," he admits. "My heart is a little heavy as I write." Mom suggests he look for a job to tide him over, but he reports no luck: "Apparently getting a job here is like anywhere else, you have to know somebody." Then Smith changes his mind and Dad can't contain himself: "Honey, it's so much like we planned that I'm a little afraid. Can it really be that we are to realize our dreams so completely?" He sends a postcard, showing the Truckee River flowing through town and promises: "We're going to live on it, and use its water to make a little lake on our ranch."

Dad always had a wonderful sense of whimsy, a quality that made him a good children's book writer and editor, and one day, when he didn't have much to do, he wandered out to a park, where a buffalo was on display for the tourists, and describes the scene: "He was about seven feet high and must have weighed a ton. Two vicious looking horns curled from his forehead. He had the nastiest look I've ever seen, and that includes humans." Dad takes a shine to the animal, names him Humphrey, and fantasizes about rescuing him from the park: "Comes a proud heir to the Rogow-Atlas gambling fortune, and Humphrey shall be his playmate and protector." Then he wonders if the huge creature is up to the task: "Humphrey comes from proud stock, but years of well-fed lolling have weakened the strain, and our baby is liable to play too rough." Of course there was no fortune, and since Mom and Dad had twins, Humphrey would have had two playmates instead of one. But even at twenty-two, Dad was eager to be a father. His affection for his unborn children seeps through the page.

So does his sense of humor. On another day in the park, he describes meeting three ducks "who strutted past me feigning indifference, but watching me through the backs of their heads. I tossed a bit of bread on the ground and they scrambled for it. . . ." But when he proffered a piece of pickle they turned up their beaks at him: "Pickle being a cultivated taste, I offered it to them again, but they sniffed, and sensing my poverty, walked away. I could see them thinking about that patronizing bug on the bench, Wait! When comes the revolution we'll sit on benches and he'll grovel in the grass while we toss him bits of live fish. Pickle, indeed!"

His overwhelming feeling is love and loneliness. He tells Mom that she'll have to travel west alone, that he can't spare the time to escort her himself, but adds: "If you arrive in the morning, we can be married before lunch." She responds to his excitement by detailing the results of a shopping spree: "I bought an inexpensive light suit to wear around (and with a view toward traveling) and I increased my nightgown purchases so I think I have now all I need." She, too, dreams of kids, but not a boy roughhousing with a buffalo: "It's so much fun buying the pretty things. How long do you think it will be before I can take our daughter shopping for a trousseau?"

She confides their secret plans to her lifelong friend Charlotte, who wants to give her a bridal shower. Mom concedes that a shower without a wedding date "would be plain bad taste," but adds wistfully: "I'd hate to miss the fun of being a bride-to-be." In the same letter, she describes seeing a cousin's newborn baby and flinching with apprehension: "I'm afraid to touch them when they're so small. What am I going to do? I could never bathe or dress or feed an infant—I'd be terrified lest I drop him or snap his arm in half. . . . I don't think I've ever *held a baby*. Do they have a school for prospective mothers in Reno? I've chewed off all my finger nails worrying about it since last night."

Through her romantic haze—and preparental anxiety—she still manages to be practical. She urges him to write home for more clothes: "Take care of it this week, Will. You'll be able to save on your laundry bill. Don't forget!" In other letters she warns him to remember Mother's Day, to avoid carrying large sums of cash around, to visit a Jewish boardinghouse in town. All their life together, she played this role, the keeper of order and security, and while Dad occasionally bridled at her needling he knew that he'd fall apart without it. "Darling, I'm afraid I'm not going to be an easy guy to keep house for," he concedes. "I have to throw my bathrobe, jacket, pants, three dirty shirts and a pile of magazines on the floor every time I sit down to write a letter. The maids here are wonderful, or I should have to cut my way out of the room with a machete each morning."

Those early days established a pattern that persisted for the next few months: delight followed by despair, fantasy by fear. He's up one day, down the next. It's painful to follow his moods, more than sixty-five years later, and it was much harder to live through them. He starts one letter: "Your husband is crying, darling . . . the deal is off." The next letter: "I don't think I'll have enough money to pay my hotel bill tomorrow." A few days later, he has a new scheme, to open his own casino, and boasts: "We're going to be rich." But back home, irritation and uneasiness are mounting. Dad's mercurial missives cross each other, confusing everybody. First Street shuts down, the family's last safety valve. Abe decides to pour his savings into an outlandish new scheme, a roadside nightclub featuring a hundred-foot tower to attract business. So when Dad asks for another $500, his father turns him down.

Mom writes that Miriam has been "rather wonderful," urging Abe to send the money "because she knows you have the right to try something on your own." But Abe is furious, sure that his son is being "hoodwinked," and Mom writes: "All those letters and telegrams have had a very bad effect on him." She, however, is furious with her future father-in-law. They were total opposites—she was as cautious as he was impulsive—and they never got along very well. She complains to Dad about his father's "whims and fancies," about his land purchases and gambling trips: "I had to exert an iron will to restrain myself from telling your father what I thought of his management." She concludes: "I was so upset I cried myself to sleep last night."

It was always painful for Dad, this tension between the two people he loved and admired the most. If his wife made life easy for him, his father made it hard. If he knew Dot loved him, he was never quite sure about Abe. If his desperate urge to please his father made him miserable, it made Mom mad. But he always tried to justify his father's erratic behavior. He explains to Mom: "When my father becomes obsessed with an idea he loses his judicious temperament. The great men in history have been the same, and I honestly believe that my father will accomplish big things, but that doesn't alleviate his impatience with me." Abe's not selfish, Dad pleads, it's just that he has "a mystical faith in his own destiny." Then he sums up his relationship to his father in a few simple, revealing words: "Just wait and see, honey. Before the summer will be over I'll be a hero to him."

One strange result of Abe's new brainstorm was that he hired Harry, Mom's father, to help him build the nightclub. My two grandfathers had a long and unhappy history—Harry once fired Abe before their children met—but Abe also knew that he needed help, that he couldn't con the building into completion. "I'm glad that my father has consulted your father about the building," Dad writes. "My father has a peculiar complex. There are many things which he does particularly well. . . . In those things, he feels that he is absolutely the best. Such things as running a stand, building, buying cheaply—well you know him. . . . But he doesn't have the thorough experience necessary for the building he contemplates. I was afraid he would try to build himself. However, by coupling my father's ingenuity with your father's experience, they ought to do pretty well." They didn't. As best I can tell, The Tower was never finished. Many years later Abe greeted news of Harry's death with smug satisfaction. He had outlived his old rival.

Dad gave Reno his best shot. The Atlas Color Wheel opened for business, and his letters are suffused with a sense that the West is different from the East, that it offers more promise and potential. The same day he writes about the ducks he enthuses: "This is a growing country with opportunity on every side as contrasted to the overdeveloped Metropolitan section." On another occasion he describes feeling "more self-sufficient" in the West than he does back home and hunts for an explanation: "Perhaps it's the prosperous environment. In the East, everyone complains and seems to find it hard to make ends meet. Out here, people own their own homes. The entire atmosphere is one of growth and development." He could have been his own father, starting out a generation earlier and comparing the New World to the Old Country, feeling "self-sufficient" for the first time and finally free of the choking, "overdeveloped" cities of Europe.

In one letter Dad describes a vigorous hike into the hills, carrying two sandwiches and a bottle of Coke that he chills in a stream and drinks with gusto. On his way out of town he'd seen a kite in a store window and bought it on impulse. After lunch he writes to Mom: "I tried to fly the kite but the wind was very erratic. A breeze would send the kite up, and I would let out a lot of line, only to have the breeze die, the kite fall, and the line [get] tangled in the brush. The kite finally broke as I pulled it along the ground."

That kite perfectly reflected his time in Reno. Breezes of hope would flare up and then always die down. The work was hard, the traffic slow, the returns meager. The future depended on chance, on the nickels and dimes played by the customers, on the turn of the wheel and where it stopped. On a good night, with enough bettors and few winners, he turned a profit. On other nights, the wheel ran "sour," hitting the wrong numbers and draining away his reserves. "I can't afford to be discouraged," he writes. "The minute I lose the drive that enables me to grind away for hours, we're finished. If I were to ease up just a little bit, everything would crash down upon us. No, darling, even when I see precious dollars carried away, even when I shout myself hoarse to an empty building, I have to have spirit to put this over." She replies: "I'm sorry I can't be stronger—that I've failed to be cheerful and encouraging when you need it so badly—when this very night you're working with bated breath, cold sweat breaking over your body as your mind rings up every dollar taken in and given out, clenching your fists as you see a dream realized and torn apart."

Abe still won't send more money, so Mom reminds Dad that she's saved $600.87 from her $21-a-week paycheck. He can have it if he needs it: "Darling, I mean it seriously—it's not my money, it's *our* money and you must use it if it will bring our happiness closer to realization." His pride is hurt but he has no choice. "I have obligations that are pressing," he explains, rent, food, bills for building the stand, "shills" who work for him and pretend to play, drawing attention and interest to the game. He remains "absolutely convinced beyond all doubt" that the game will work, but Smith wants his money for the space, $9 a night, and Dad sees that the corners of their future are starting to crumble: "With this intense strain, tonight was a nightmare. No one could lose. . . . I think I might have had a tear in my eye as I saw my bankroll diminish to the point where it was impossible to stay open any longer, for fear of getting a hit I couldn't pay off. I haven't the money for Smith tomorrow and I haven't enough to open up on."

Finally, he broke down. He sent Mom a telegram, asking for money. Then he wrote two bad checks, one to Smith for $9, another for $4 in pocket money. His bank account wouldn't cover either one. He opened the stand for business that evening, knowing that he had to produce $13 by the next morning. But the winners kept coming. "Dotty," he wrote, "I felt like a whipped dog, I couldn't get ahead." After ending the night even, he went back to the hotel but slept little. "The bank opens at ten, and I knew the two checks would be presented shortly after," he wrote later. "If your money order wouldn't come before then, I'm afraid to think what might have happened." But at nine, the money order was slipped under his door. He was saved for another day: "I began to laugh hysterically."

It was a temporary reprieve. Back in Bayonne the Rogows were heading for disaster: Labor trouble closed down work on The Tower; political trouble kept First Street shuttered; creditors grew more demanding. The only family income was a sorry little hot dog stand Abe had opened. Miriam summoned Mom for a frank talk. Will would always have a place with his family when he returns from Reno, she said, but if he gets married, that would change things. There was no room for a wife.

Mom finally confronted the facts. She would not quit her job, she wrote, her only source of security, and she would not be coming west: "We've got to stop being such god-damn fools. Even if you make hundreds during the summer—what are your chances of making nickels and dimes during the winter? And the winter will come." Perhaps she should have been the writer in the family, her letters were always crisper and more cogent than his, and now that she's into truth-telling, she doesn't stop: "Will, dearest, I know how it hurts you when I criticize your father—but this much I must say—I don't want you to work for him if we are to be married. I don't want my life and happiness and security at the mercy of his whims and caprices. . . . Perhaps, as you think, your father is an empire-builder. But if I want any empires, I build them myself, the way I want them."

It's in this period that she sends the letter I quoted from earlier, urging him to focus on writing as a career, not gambling: "Write until you're blue in the face, even if it's only to tear it up when it's finished. Out of a thousand words, ten may be good, out of ten thousand, a hundred. But none of it is any good as long as it's just mental gymnastics." She admits that "it's not easy to sit down to write when your mind is the breeding place of a thousand worries and anxieties." But she wants him to follow his "natural inclination toward a life's work," and she's willing to accept the financial sacrifices that might require. "As long as I know you're doing the work you were meant to do, the work you want to do, I'll find strength for the both of us to wait. . . . Perhaps after a while, when your articles begin to pay, and if the Street opens and you can work there nights, we'll be able to see our way clear."

He responds with excuses: "Darling, I too have been realizing increasingly the importance of doing something about my writing, even though I doubt if I can do much in Reno. Yet making a living from writing is another thing. . . ." *The Nation* barely pays its contributors, he notes, and most have to earn outside income as teachers or reporters for other publications. Getting published might lead to growing "recognition," and even "a job with a larger publication." But he warns her: "This is a long arduous process hardly to be relied on." He makes her a promise: "I'm going to write," not just critical articles but "more marketable fiction" as well. Right now, however, he can't really think about that. "The problem of making a living still imposes itself."

Finally, it's clear to both of them that his problem of making a living won't be solved in Reno. "I wonder how I could have been so blind," she writes, "this has all been a terrible mistake, planning on being together in a few weeks." Then she answers her own question: "I loved you and wanted you and shut my eyes to everything else." His final hope, a big July Fourth weekend, doesn't pan out. He makes enough to pay his bills, but that's all. All the money they'd invested, including Mom's, is gone. On July 9, he writes his final letter from Reno, the last one from Will Atlas in room 267 of the Hotel Golden: "Darling, I'm coming home. My bus leaves in just a few minutes, so I haven't time to go into detail. The choice isn't mine (Smith was insistent) but quite agreeable with me." Within a few weeks, he concedes, "I would be left penniless and in debt, and would have to wire home for bus fare." He was saved that final humiliation, but just barely. He writes from Omaha that he should be home soon: "I'll wire my father exactly when to meet me and if I have 35 cents left I'll wire you, too."

Today, Mom remembers what Dad looked like as he stepped off the bus: "He was so emaciated, he hadn't been eating properly, if at all. He was green. He didn't have enough money to stop on the road so he took the bus all night long. He was so defeated. It was pathetic." But not completely defeated. Their dream was not dead. They still wanted to be together, and less than a year later, they found a way.

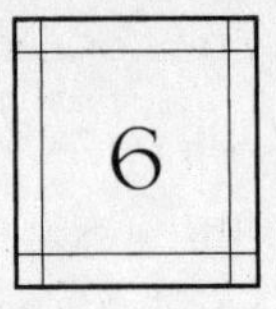

THE "TWINNIES" ON THE BLOCK

The pipe dream of a "Rogow-Atlas gambling empire" died in the western desert. There would be no ranch, no lake, no ducks to feed on bits of bread and pickle. Humphrey the Buffalo would never know the pleasure of romping with the Rogow offspring. Once back in Bayonne, Dad had to move in again with his parents. Their place was small and overcrowded, and Abe and Miriam's only income came from the hot dog stand on the highway, a shack Mom describes as "pathetic" and "degrading." Harry's house was also jammed, since Mom's older sister, Sylvia, had gotten married and brought her new husband home to Thirty-first Street. Dad got a job selling advertising space in a magazine, but commissions were sparse. Mom kept her job at Maiden Form, even though she hated every minute of it. She couldn't turn her back on $21 a week, their only regular income. As Mom described that time: "There wasn't any elbow room to spread your wings or dream about what the future could be. Nobody knew what it was going to be."

But Will and Dot were still in love, and love finds its ways. "We just decided we were going to get married," Mom remembers. "Period." They had to keep it a secret, however: "We couldn't afford to live together. We had no place to go." So on May 27, 1940, about ten months after Dad's return from Reno, they carried out their plan. On the appointed morning, Mom walked the block to Dad's house and ironed a fresh shirt for him to wear. Then they took the bus into New York—she was decked out in a navy blue suit and a red blouse—and went to City Hall. To protect their secret they registered using false addresses in Manhattan. Their only attendant was a man named Harvey Bond, the boss at the magazine where Dad was selling advertising. Bond brought a gardenia for the bride. Dad carried a heavy metal coin bank, which contained the money he had saved for a wedding lunch, but midway through the brief ceremony, the bank slipped and crashed to the floor. Fortunately, it didn't break. Afterward, they had lunch at a Bickford's cafeteria and spent the afternoon seeing the show at the Radio City Music Hall. They couldn't afford tickets, but Dad had a college friend whose father directed the orchestra, and he finagled them free passes. Then they returned to Bayonne, back to their separate houses. It's a sad story in one way, very romantic in another. They were apart, and they were poor, but they were married. "It's what we had been dreaming about for so long," recalls Mom.

Finally, fate tossed them a bouquet. Sylvia and her husband had been saving their money and a few months later found their own place. Mom and Dad immediately announced that they were, in fact, married. I doubt if the news surprised anybody, since they had been talking about marriage for so long. The day Sylvia decamped, Dad moved in—and stayed for the next sixteen years. Mom had missed the joys of being a bride-to-be, joys she had eagerly contemplated in her letters to Reno, but after she and Dad went public her girlfriends staged a celebratory luncheon in New York. I've seen a picture of that event, a dozen young women gathered on the sidewalk outside the restaurant, happy and elegant, as they feted their friend. And not one of them was carrying a metal coin bank. Then fate snatched the bouquet back. Dad was hit by a car while crossing a street in Bayonne. His pelvis was broken, and he was placed in a full body cast for a recovery that would take months. Since Mom had to work, she couldn't take care of him during the day. So once again, he moved back in with his parents. A photo from that time survives, the only one I've ever seen of Dad wearing a full beard.

Once Dad mended, he went looking for work, but the Depression still dominated their days. "He had the stereotype of being the wage earner and he wasn't earning any wages," Mom recalls. "I don't think he knew where to turn, what direction to go in." But she knew one direction he should not go in, back into the gambling business. "I felt that his continued association with gambling was really beneath his abilities. It was too much of a compromise. He had to get rid of it, he had to shed it, once and for all." Finally, Dad got a job as a stock clerk at a New York department store, Oppenheim Collins. While lunch and carfare wiped out most of his paltry salary, the stint left Dad with a very useful talent: He could make the best packages I've ever seen. In fact, he used to joke that he was a member of the Shipping Clerk's Hall of Fame. He once sent our daughter a tennis racket that must have taken him hours to wrap. After receiving it, she remarked that the package could have dropped directly from the plane and still arrived intact.

In mid-1942, Mom became pregnant and immediately quit work. In later years she sometimes kept the books for Dad, but never held a paying job again. As she had written, she wanted to be considered an equal, but had no interest in acting on that equality. I believe her when she says: "I was content to be an old-fashioned wife and mother. I was who I was and that was it." Mom didn't know she was carrying twins and we arrived early. Dad was working in New York when he got word that her labor had started, so he rushed back to Bayonne. Told it would be a while, he went out and stocked up on magazines and pipe tobacco and arrived back at the hospital, ready for a lengthy siege in the waiting room. When he walked through the door—it was February 11, 1943—the doctor delivered the unsettling news: He was already a father. Of twin boys. Here's how Mom described the scene in our baby book: "The new Daddy didn't believe it was twins and when Dr. Lipschitz finally convinced him, he went off in a corner and had a good laugh at his own expense (and we do mean 'expense')." She added: "Grandpa Rogow was a good second-guesser—he knew it would be two all along." Yeah right, Pop. As for Mom, when she recovered from the anesthesia enough to hear the news, she murmured, "What a lot of baby," and went back to sleep. Miriam dubbed us the "twinnies."

By her own admission, Mom was inexperienced in matters maternal, and the absence of her own mother, who had died four years before, aggravated her anxiety. The vogue at the time was a very controlled feeding regimen, and we cried all the time, apparently because we were starving. But we survived anyway, and my first words went down in family lore. I was playing in the bath, happily splashing the water, and said, "Goddamn it." I'm sure it was a direct quote from my father. So right from the beginning he was a prime influence on my use of words.

Being a twin is a strange and special way to grow up. You are seldom alone. Most of our baby pictures show us together: in a carriage, in a sandbox, dumping blocks on the floor. Looking back, I think this was a good thing. A twin never labors under the illusion that he or she is the sole center of the universe. In my entire life, I've had my own living quarters for exactly one year. And to this day I can't stand celebrating my birthday alone. I have two nephews with similar birthdays, and for years we've had a "three cake" party, where we all join together. And my oldest grandchild, Regan, got herself born on February 11 as well. Obviously a clever child. So even when my twin is not around, I'll always have a birthday buddy.

Sure, there were times when Marc and I hated each other—he split my head open at least twice, once with a dried corncob, once with a board that had a rusty nail protruding from it—but we always had a companion and a playmate. (Although my mother does recall the time we appeared at her feet and proclaimed, "We don't have anybody to play with.") We quickly learned, as all twins do, that acts of mischief that seem impossible for one child can easily be accomplished with some twin teamwork. Since this was during the war, rationing made many foodstuffs hard to get, and Mom had been hoarding packets of Jell-O for a special occasion. She woke up one morning to find that her darling boys had stacked enough chairs or boxes on top of each other to reach the food cabinets, where the Jell-O was stored. Then we had proceeded to empty every single precious packet, dissolve the brightly colored powder in water, and wash it down the drain.

There's another story Mom tells that recently acquired new meaning for me. We lived on the second floor of the house on Thirty-first Street, but one morning—still dressed in our pajamas with feet attached—we managed to go down the stairs, open the front door, and toddle up the block. Mom was startled by a call from a neighbor saying: "Dotty, do you knows the boys are out on the street?" The new meaning comes from my twin grandsons, Cal and Jack, who at age two and a half decided they would go looking for Sir Toppenhat, a character who runs a railroad in their favorite series of stories. If they found him, they were convinced, he would give them each a new train. So they went to the front door, put on the only shoes they could manage on their own—red rubber boots—and marched out on their mission. (Cal neglected to put on pants, but he remembered the boots.) When my daughter realized what they were up to, she did the only sensible thing—grabbed the movie camera and followed them. Twins will be twins.

When I'm playing with the twins or just hearing stories about them, I feel like I'm reliving my own childhood. One night Marc and I figured out how to climb up on the chest near our cribs, and devised a game of leaping happily back into our beds until we were interrupted by an "absolutely hysterical" mother, in Marc's words. When Cal and Jack visit us, their cribs are arranged against a wall that is lined with a bookcase. Their mother discovered them one morning in the same crib. Jack had figured out how to climb along the top of the bookshelf and launch himself into his brother's bed. It was not the first, or the last, time I was reminded of a saying, coined by my sister-in-law's mother-in-law. When her children would complain about their own offspring, she would calmly observe, "They don't take after strangers."

Grandchildren connect me to my father in an almost mystical way. When I'm physically close to those little ones, in a bath or a pool or just reading in a chair, I can sometimes sense my own father's arms around me, caring and comforting. He was not a particularly tall man, a few inches under six feet, but he carried a few extra pounds around, and his physical presence radiated an aura of solidity and security. While he never saw any of his great-grandchildren, I feel he is reaching and reassuring them through me. When I was very young, I had a birthmark removed from my shoulder, and the doctors told my parents to check the scar regularly, to make sure nothing was growing back. Dad would run his fingers lightly over that scar, a particularly intimate and loving gesture. When I hug my own grandchildren, I hope I'm able to convey the same sense of strength and safety I felt from Dad's touch.

I remember not just his touch, but his things, the smell and feel of them. Watching him fuss with his pipe—filling it, cleaning it, puffing it, pointing it—was a constant part of being around him as a child, and the aroma of his favorite tobacco, Walnut, which came in a distinctive white tin, permeated the household. When I joined the *New York Times*, I loved being around pipe people—my boss, Scotty Reston, chief among them—because every wisp of smoke or cascade of ash reminded me of Dad. I always regretted the decline of pipe smoking because it snapped a link in my chain of memory. While I worked for Scotty, Dad even sent him a can of Walnut as a present, so for a time, until the can ran out, my father and my mentor actually smelled the same. Then there were Dad's shoes. He wore exactly one style of shoe from the same maker my entire childhood, while filling me with fear that ill-fitting footwear could be my ruination, a vice as lethal as alcohol or slothfulness. His prohibition included just about any style without laces, and I was deep into adulthood before I was able to overcome his admonitions and actually purchase a pair of loafers. All kids rummage around in their parents' closets, and I would always come upon three or four identical pairs of shoes lined up on the floor—Florsheims, sort of medium brown, with pebble-grain leather and plenty of laces, all in different stages of wear. All my life, whenever I passed a Florsheim store, I thought of Dad, and when the company closed its stores a few years ago, I felt a sharp stab of regret. No Florsheim signs meant one less reminder of the man who influenced my life more than anyone else. Recently, much to my delight, a few Florsheim outlets have reopened. Now, if only I could find a co-worker who smokes a pipe and uses Walnut tobacco.

Dad loved being a father, it brought out all the playfulness he occasionally revealed in his letters from Reno, and by the time we came along, some of the financial tension that had plagued him for so long was starting to ease. He had gotten his first real job, at Sloves Mechanical Bindery in New York, but he was able to set his own work hours and was around a lot during the day. There was a park a few blocks away, and Dad took us there often. In good weather we'd climb on the jungle gym or swing on the rings, and I can still recall the odor of cool water splashing onto the hot metal of the drinking fountain. In winter he took us sledding on Double Hill, a slope divided by a walking path, and on a really good ride you could reach the edge of the woods. To this day, the smell of

Me, Dad, and Marc clowning around on the steps of Abe and Miriam's house on Lincoln Parkway. We were probably escaping Abe's ritualistic devotion to Meet the Press *on a Sunday morning.*

wet wool drying on a radiator brings back those outings. But we were not always easy children and he was not always a patient parent. When we got rambunctious and tried his temper, he would usually come out with a favorite saying: "I don't care who started it, I'm stopping it!" I found myself using the same phrase with my own kids more than once. One day Dad, never much of an athlete himself, took us to the park and was teaching us to throw a ball. When I couldn't get the hang of it right away, I turned cranky. So he simply packed me up, brought me back home, and returned to the park with Marc.

The incident became a watchword in our household: Don't behave like Steven when he couldn't throw the ball.

All families have their own folklore, stories they tell and axioms they use, and I contributed one the day Dad took Marc and me into New York to visit his office. At lunch he ordered us soup, and I was fascinated by the small packets of crackers that came with my meal. I opened them all up, crumbled the crackers into the soup, and asked for everybody else's packets as well. Then I refused to eat. Why? Dad asked. "Too many crackers," I replied. For years afterward, whenever someone in the family didn't like something, Dad would invariably say, "Too many crackers." Going to Dad's office in New York was one of our greatest treats. He rented several lofts in Lower Manhattan over the years, and books were stacked everywhere in those vast open spaces. There was a Horn & Hardart cafeteria near one of the lofts that featured a special gimmick: The food was displayed behind a glass window and when you put in the right number of nickels the window opened and you took out your selection. It was a big deal when we were old enough to put the nickels in and extract the egg salad sandwich or apple pie by ourselves. At holiday time, when Dad had large orders to fill, Marc and I would ride on top of the pushcarts, stacked high with packages, as we headed through the crowded streets to the post office. Apparently we would inform anybody who cared to listen, "This is my Daddy's stuff." True enough. But it wasn't just any old stuff. It was books. We learned pretty early that they were special.

Grandpa Harry quietly accepted a new son-in-law into his house, pipe ashes and all, and then two hyperactive grandsons. A man with a fiery temper when he was younger, Harry turned so meek in later years that Marc describes him as a "ghost," a ghost with a full head of white hair and a small, sweet smile. He owned the house, but rarely asserted his authority. The refrigerator occasionally contained two favorite dishes from his Russian past: borscht, a dark-red beet soup, and something called "schav," a weird green concoction made from leeks that tasted as bad as it looked. Those were virtually the only signs of his presence. Occasionally, I wandered into his room when he was out, and I recall a slightly musty smell I've associated with old people ever since. He was always kind to the "twinnies" and often took care of us when Mom and Dad went out. I drove him crazy by insisting that he scratch my back, but he never got angry, he'd just gently push me away. I never saw his famous capacity for fury, but he did preserve another trait from his youth, leftist politics. He became a loyal member of the Workmen's Circle, an organization fostering Jewish culture and progressive ideas, and he'd take us occasionally to the Labor Lyceum, a sort of clubhouse for immigrant Jewish socialists. I liked going, but not for the politics. There was a big jar of free pretzels on the front counter. One of my passions as a small child was collecting chestnuts from the backyard of a house we passed on the way to school, and Harry noticed my interest in those small brown trophies, which I kept in a dresser drawer. He often visited a Workmen's Circle home in the nearby town of Elizabeth, which was surrounded by chestnut trees, and in attempt to be a good grandfather, he once brought me back a whole bag of chestnuts. How could he know that he was ruining the whole game?

Between Harry and Miriam I don't think we ever had a babysitter who wasn't related to us, and I thought everybody grew up that way, with grandparents in the neighborhood who were always a part of your life. I learned differently when I left Bayonne, of course, but when we lived in Greece many years later, I was delighted to learn that the Greek word for babysitter is "babysitter." There was no original word in the language because the concept was so foreign. Children were minded by their relatives, not strangers. Just like on The Block. And that's still true.

Some years ago, when I returned to Bayonne on an assignment for *U.S. News*, I discovered twenty members of one extended family, the Donnellys, all living within a few doors of each other on West Thirty-first Street. The three Donnelly girls, all nurses, had married men named Pawlowski, Amato, and Seestead—mixed marriages, Bayonne style. One night Denise Pawlowski was babysitting for friends who had to attend a wake while husband Randy was caring for the three blond Pawlowski boys. When the youngest, three-year-old Ricky, managed to get something stuck in his ear, the family swung into action. Nancy Amato relieved sister Denise, Anthony Amato took over for Randy, and the Pawlowski parents rushed Ricky to the hospital. By the next night, all was calm and I talked to Randy Pawlowski, a mail carrier, about life on The Block. It gets "very hectic, very hectic," he admitted, with lots of kids and little privacy. But as the previous night demonstrated, the benefits are substantial: "If I was out in Middletown somewhere, what would I do? Ask some stranger to sit with my kids?"

Harry and his son-in-law did have one run-in that showed a rather absentminded side to Dad, a man who ran out of gas more times than I care to remember. When we were small, and my parents did not have other plans, they would sometimes order delicatessen sandwiches from Botwinick's, a local restaurant, as a Saturday night treat. One night Dad asked to borrow Harry's truck to collect the sandwiches. He came home some time later and announced, with considerable embarrassment, that he had locked the keys in the truck. Could he now borrow Harry's car to fetch dinner? But then he did it again, this time locking the keys in the car. "My father blew his stack," Mom recalled. No kidding.

Most of my early childhood is a series of snapshots, pasted randomly in a mental scrapbook. Dad couldn't sing a note, but he could whistle, and he would put us to bed by ranging through a repertoire that featured Broadway show tunes. *Oklahoma!*, which opened the year we were born, was a particular favorite, and every time I hear "Surrey With the Fringe on Top," I can almost feel Dad patting me on the bottom to the rhythm of the song as I fell asleep. On our fourth birthday, Mom had hidden our cake out of harm's way, and when our friend from down the block, Margo Metzger, came early to the party, we just had to show it to her. So, of course, we dumped it on the floor. When our brother Glenn was born, six months after the cake caper, we were moved to a new room and given "big boy" beds. We started jumping on our new furniture and announced we were going to laugh until Dad came home. I doubt if we made it.

As our economic situation improved, we rented a house in the New Jersey countryside one summer, and I can still smell the wild onions growing next to a stream where we used to take picnics. To this day, I hear the word "picnic" and I immediately recall those onions. One morning, when I was about four, I reported to my parents that there were "horsies on the lawn." Everybody thought I was making it up—a very young writer already spinning tales—but when I persisted, the grown-ups looked for themselves. Sure enough, there were horsies on the lawn (the boy became a reporter after all, not a novelist). They had escaped from a nearby camp and were enjoying breakfast in our yard. When Dad drove to the camp, to tell the staff where to find their horses, we went with him and got free rides as a reward for our vigilance.

Except for those brief sojourns to the country, The Block, West Thirty-first Street between Avenue A and Newark Bay, was our whole world as small children. (Bayonne's founding fathers showed a distinct lack of imagination when it came to street names, since most of the east–west streets were numbered, and the north–south avenues were lettered.) When I once suggested to my brother Marc that Bayonne was a pretty isolated place, and that New York, or even Jersey City, was another country, he replied: What are you talking about?

Thirty-second Street was another country! By the time I was born on The Block, my mother had already lived there almost twenty years, and it seemed like a timeless, unchanging place. The houses were almost identical—two-story, two-family dwellings, built right on the street and right next to each other. Few families moved in or out, and almost everyone I knew had a grandparent living with them, usually with a fragile grip on the English language. That immigrant tradition continues today in Bayonne, where one out of five residents is foreign born and one out of three speaks a foreign language at home. The city is not as insular as it once was. Ten Bayonne residents were killed on 9/11, and most worked on Wall Street. Yet on a recent visit, The Block looked very much like it did half a century ago. Our house had been spruced up, with yellow aluminum siding and an American flag flying from a pole in the tiny front yard. But the whole feel of the place was very familiar. I was particularly pleased to see a basketball goal in the backyard, since I spent a good part of my childhood shooting hoops in that very spot.

During my return to The Block for *U.S. News*, I approached a man who was hosing down the sidewalk directly across from my old house and asked to talk. He wasn't interested, or even very cordial, but he suggested I consult his wife, who was at home in the second-floor apartment. She wouldn't open the door at first, but agreed to listen through the glass. When I told her my name, and that I used to live across the street, she said immediately through the door, "And your mother is Dorothy and your father is Will and your brother is Marc . . ." She invited me upstairs and, as we talked in her living room, I could see across the narrow street and into the front room of our old house. That room had been a little sunporch when I was growing up, but it was a small house and we only occupied the second floor, so when my sister was born when I was ten, the porch became her bedroom. In fact, just about every space in that house eventually became a bedroom.

An hour later, I stopped to talk to a woman near the corner of Avenue A. She was standing on the street, leaning against a car, and when I told her my name she said in horror, "Steve, you don't remember me?" Frantically I tried to count down the houses from my old address. "Of course," I said in triumph, "Mrs. Fedorochko!" But that only made it worse. "Steve," she said indignantly, pointing to the woman sitting in the car, "this is Phyllis." Then I knew. It was Mrs. Cizik. I had gone to school with Phyllis Cizik. I had been off by one house. One house, after thirty-five years, and she was still steamed at me. But that's Bayonne. Everybody has a place, and you're expected to remember who fits where.

Margo Metzger and Phyllis Cizik were only two of the many kids who lived on The Block. Almost every house had them, and since the street was a dead end, it was a perfect playground for small children. One of our favorite holidays was Halloween, when we would all dress up in costumes and troop from house to house, asking for candy. The best target, however, was the woman who gave out nickels instead of Tootsie Rolls. The really adventuresome kids—a group that

A circus in the backyard of our house on Thirty-first Street. I'm the cowboy on the far left in the striped shirt that won't stay buttoned.

did not include me—would go all the way to the adjoining blocks and really haul in the loot. I have a picture from Halloween of 1947, when I was four, taken in our backyard. Marc is dressed as a cowboy and looks quite sharp; I have on a droopy clown suit that makes me look like a dork, and I don't appear all that happy. That backyard also contained a sandbox, and we used to invite the neighborhood kids over to play. But Margo complained that we always threw sand in her hair.

I had a lot of fears as a child, and Dad always blamed a doctor who rudely and painfully pulled out a splinter when I was very small. I'm not sure if his diagnosis was correct, but I was petrified one day when one of our neighbors, a beefy Polish boy named Peter Polakowski, told me there was a "Peeping Tom" on the block. I had no idea what a Peeping Tom was, but I was sure he wasn't friendly. In our backyard was a tall wooden pole, perhaps twenty-five feet from the house, and a clothesline on a pulley system was strung from our kitchen window to the pole. (The first picture in our baby book shows that line full of drying diapers, a terrifying sight.) For years, once evening fell, I could not look out the back window, convinced a Peeping Tom had climbed the pole and was intent on somehow peering into our kitchen. Perhaps he was looking for the lost Jell-O.

During the day, though, The Block felt like a totally safe place. Even when we were quite small, Mom would toss money down from our second-story window, wrapped in a shopping list, and send us on simple errands around the corner—to the butcher, or the bakery, or Levine's Fish & Vegetables Free Delivery, as the sign on the window advertised. At the grocery store, the clerk used a long pole with a gripping mechanism on the end to pull down boxes of cereal or soap powder from the top shelves and totaled up your bill in thick pencil on a brown paper bag. It was also a time when many products and services were delivered to your door. Coal trucks would stop on the street and laborers would carry the chunky black fuel in filthy sacks, slung across their backs, and pour it down a chute into the bin in the basement (a favorite if somewhat scary place during games of hide-and-seek).

Seltzer and milk also showed up regularly. But in warm weather our favorite was the Good Humor truck. When we heard the distinctive bell, signaling the truck's approach, we'd set up a cry for money and moms all over The Block would toss coins down to their impatient offspring. One of my favorites was a stick of vanilla ice cream coated with orange ice, a Dreamsicle. To this day, when I see a combination of orange and white—say, in a flower bed of impatiens—I think of that frosty delicacy. A rival brand would occasionally try to intrude on The Block, to snare a share of the Dreamsicle market, but we were having none of it. We stayed loyal to the Good Humor Man. That's just what you did in Bayonne.

WARS AND WORDS

The war years finally ended the Depression for my family. They also encouraged the careers of both my father and my uncle Bussy, two men who had such a large impact on my writing life. While I was still playing in the sandbox in our backyard, they were pioneering the paths I would later follow. But the most immediate impact of the war was commercial, not creative. In early 1941, the U.S. Navy took over a large port terminal on Bayonne's east side that jutted into New York Bay. After Pearl Harbor the Navy tripled the size of the terminal, and Bayonne became the main supply point for naval forces in the Atlantic. Suddenly jobs were plentiful and by July 1943, when I was five months old, the city had erected a large billboard that drivers coming across the Bayonne Bridge from Staten Island couldn't miss: WELCOME TO BAYONNE. A CITY OF WAR WORKERS. The message: Don't make noise because shift workers could be sleeping in the middle of the day. Mom was grateful, she didn't want people waking her twins in the middle of the day, either. Other Bayonne businesses were booming as well. Maiden Form adapted its bra designs to fashion special vests for paratroopers. The pouches held messenger pigeons as the soldiers dropped from the sky. The Elco Naval Division built the famous PT boats not far from where we lived along Newark Bay, and the "mosquito fleet," as they were called, often did test runs in the waters off Bayonne. Jack Kennedy's PT-109 might have scooted right past Thirty-first Street while I was sunning in my carriage.

Grandpa Abe took a job at the naval base as a carpenter. One of his assignments was to remove sheets of plywood that had been used as forms for poured concrete. Abe noticed the used boards were being discarded, and asked if he could have them. When the Navy agreed, he bought an old truck and carted the wood away. In typical Abe fashion, he found a market in the southern Jersey town of Lakewood, where a colony of Russian Jews had started a number of chicken farms, partly as a way of obtaining agricultural deferments from the draft. The farmers used the old forms to build coops, and Abe became friendly with several of them. One day, when I was about two, we took a family outing to Lakewood and visited one of the chicken farms. I got cornered by an angry rooster, who ran back and forth, preventing me from leaving the yard. Instead of rescuing his weeping child, Dad took home movies of this embarrassing event, which quickly became a staple of family folklore. In fact, the movies were shown at my rehearsal dinner, the night before my wedding. Thanks, Dad. Many years later, while I was a foreign correspondent based in Athens, I was chatting with a State Department officer named Ray Benson on a snowy street corner in Ankara, Turkey. He mentioned he was from Lakewood, I mentioned that my folks had moved there a few years before. The coincidences kept building until I realized that Ray's family had owned the chicken farm that was the scene of my youthful humiliation.

My aunt Rose also got a job at the naval base, bringing home both a hefty paycheck and a young officer, Hy Mishlove, who loved Miriam's home cooking and later married her daughter. Dad's cousin Pearl Weiss, an orphan who lived with the Rogows, worked at the base as well and also married a Navy man. Some time later her husband, Mike Bronstein, was aboard a ship that sank in the Mediterranean. He was brought to a hospital in Europe, along with other survivors, and one day a nurse walked into his room and said, "Hi, Mike." It was his wife's sister. By complete coincidence she was stationed at the hospital. After the war many naval vessels were mothballed at the Bayonne base, and once a year, on Armed Forces Day, local Boy Scouts were allowed to climb all over the ships and even spend a night in their cramped and crowded bunks, stacked in tiers of four. It gave me a tiny sense of what Hy and Mike and so many other relatives had lived through during the war.

All around us on Thirty-first Street, men were leaving for

the service. Every house proudly displayed pictures of soldiers in uniform. Our neighbor, Marv Simon, wound up fighting in the Battle of the Bulge, and while I was never quite sure what that was, I somehow knew that he was a hero, that he'd seen and done things the rest of us never would. Three of my uncles wound up in the service, and I have a photo of my uncle Murray, in his new Coast Guard uniform, peering into our carriage with the Newark Bay in the background. My own father never served. He had a trick shoulder, courtesy of that youthful fistfight on First Street. And when he deliberately threw the shoulder out of joint during his medical exam, the military rejected him as physically unfit. I understand why he didn't want to go. He already had two children by the time the draft board called. And I'm making no moral judgments,

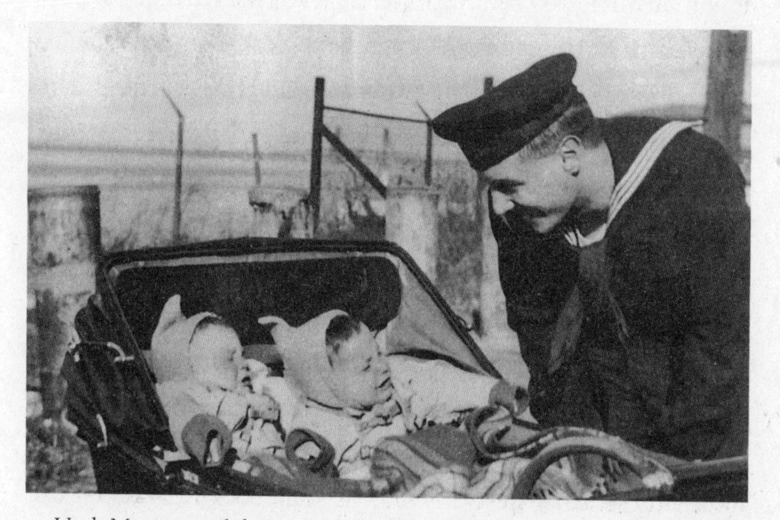

Uncle Murray, my father's younger brother, wearing his Coast Guard uniform and frightening his twin nephews. During my childhood, every house on Thirty-first Street displayed pictures of men in uniform.

because I used similar tactics to evade the draft during Vietnam. But it must have felt strange for him, staying home while most men his age were away.

One of them was Dad's brother Lee, Uncle Bussy. By his own admission, Lee was a reluctant warrior, and in a journal he started on the day of his induction in early 1942, he wrote: "It should be said right now that I have never wanted to be in the armed forces, and I would not be in them right now if I could see any reasonable way of avoiding service." Like Dad, he'd been a "heavy antiwar movement agitator" during his days at NYU and a professed "isolationist." But just months after Pearl Harbor he now admitted: "This was purely wishful thinking, of course. I merely hoped we would never become involved in a war, because I liked the life I was leading too much to want it to be interrupted." That life included writing advertising copy for an agency in New York; composing musical shows at Grossinger's, a Jewish resort in the Catskills; and romancing glamorous women, including a singer named Sugar who was then his girlfriend. But he didn't have a trick shoulder, or twin sons, and when Abe and Rose got jobs at the naval base, he could no longer claim that he was supporting his family. He learned about Pearl Harbor while lazing around Grossinger's on that infamous Sunday morning and knew immediately that his life was about to change. "I remember the ride home from Grossinger's," he wrote in his journal. "I sat in the front seat of somebody's darkened car, and the world spun dizzily inside my head. 'This is it,' I kept saying over and over again to myself. 'This is it.'" His first impulse was "to love and be loved, to get drunk, to go places I had never been, to do everything I had been promising myself I would do." But his options for revelry on a Sunday night in December were limited. He settled for buying a big chocolate layer cake at Cushman's Bakery in Greenwich Village. "I think I was almost gay as I ate the cake that evening," Lee wrote. "It was a good cake."

Some weeks later he was accepted into the Naval Reserve as a "ninety-day wonder," and became an officer after only three months of intensive training. For that period he lived in a dorm at Columbia University—the first Rogow to make it to the Ivy League—and learned sailing in the Hudson River, not far from the naval base where his father and sister were working. Training lasted from six in the morning until ten at night, a brutal schedule, but Lee still found time to write, and not only in his journal. He became the editor of a 168-page book published in August 1942, chronicling the experience of the ninety-day wonders. The family always got a big kick out of the page listing the staff. There was Lee's name at the top in big letters. Down on the bottom in small type was a classmate identified as "H. Wouk." That would be Herman Wouk, who later did okay for himself in the writing business.

Military service made a big impact on Lee, maturing him as a person and sharpening his skills as a writer. He was clearly referring to himself when he described the training regimen in his journal and added: "Ordinarily, you'd expect a schedule like that to produce a large amount of grousing from healthy American boys, most of whom got through college majoring in saddle shoes and chocolate malteds. But, strangely, you hear none of that." He offered two explanations: "First, this is really fun. . . . Second, it all has a point." He and his 789 classmates were being trained to take "direct action in the conflict which is now the single most important thing that has happened in our lifetime." He wasn't composing songs or courting singers anymore. The skills he was learning would determine "the safety of [my] ship and the lives of the people on it."

In his introduction to the book, Lee expresses his growing feelings of patriotism as he describes his classmates: "They have become Navy men because America needs Navy men." But he can't suppress his natural instincts for humor, either. Writing about the electrical engineers, he says: "It all started with Michael Farraday and trouble-makers like him." He describes the complicated process Farraday invented to "generate a peculiar form of energy known as electricity." An unfortunate development, Lee concludes: "If it were not for this initial mistake, all the boys in the 'Juice' course could sleep." Then there were the signal specialists: "The Deck men got so they could read blinker like a native tongue, and many the happy hour they spent sending obscene words to each other in this delightfully private lingo." For a young man raised in a relentlessly secular household, his reaction to the weekly chapel service is revealing: "For here, each Sunday evening, each man found himself in the midst of an intensely personal experience. It was a surprise to most of us, many of whom had lost the habit, and the feeling, of 'going to church.'" But Bussy is at his best writing a column called Beautiful Blunders. He recalls one midshipman who attended a fancy dance in New York and urged a young lady to exit through a set of French windows and join him on the terrace. "With a flourish," Lee writes, "he swung one of the windows open and stepped aside, like a perfect little gentleman, to let the lady go through first. She smiled encouragingly, picked up her skirts, stepped over the threshold, and plummeted two stories to the garden beneath, fetching up smartly in a rose bush. No terrace."

Bussy's journal continues after his commission, while he's serving on a naval vessel escorting a convoy of tankers headed for Iceland. He writes about the books he's reading, the poker he's playing, the occasional alarms that punctuate long days of boredom. He writes about Sugar: "I remember you in that white terry cloth robe, and how your face looked up at me when I kissed you last." He lists the publications he's edited, including the Bayonne High newspaper, the *Beacon*, which I would also run some years later. And the girls he's loved, starting with "plain little Carolyn," when he was eleven. Clearly he was not as "chaste and virginal" as his older brother. He does allow that the publishers of Casanova's memoirs "need have no fear that these intimate revelations of the Lothario Lee will drive their volume off the best-seller lists." But he's no frightened boy, whistling in the dark, when it comes to women. And he flavors every experience with his sense of humor. In the middle of a huge storm at sea, he recalls this scene: "I approached each of the other men quite seriously, as we braced ourselves against the tossing of the ship, [and] I would ask with a straight face, 'Have you noticed that the ship seems to be rolling from side to side a little as a result of the action of the sea?'" A few years later, Lee was training to command a landing craft for a possible invasion of Japan. It would have been a suicide mission, but President Truman dropped the atomic bomb, killing a lot of Japanese and saving a lot of Americans, including my uncle. No one in the family ever had any doubts that Truman made the right decision. Lee survived the war, got married, started a family, and launched a luminous writing career that was still climbing sharply when his plane fell from the sky in September 1955.

The war directed Dad's life down a different road, but it still led him into the family business: Rogow's Writers. Words and Ideas for All Occasions. During his tenure at Sloves bindery, he started working with the customers—authors, editors, publishers—and learning about the book business. He still lacked the self-discipline to write serious articles for *The Nation*, or the light verse and humorous sketches his brother produced by the barrel, but he discovered that he had other talents. He could conceive of a book or a toy for children and translate that idea into reality. Mom says that being a father made a big difference: "It was the association with you boys that stimulated his imagination. It was a magical time for him."

As a member of the Shipping Clerk's Hall of Fame, Dad

was always handy with a knife, one knife in particular. It was called an X-Acto. It was metallic gray in color, fit easily into the palm, and contained a disposable, razor-sharp blade. If I associate Grandpa Harry with a wooden toolbox, I think of Dad with one of those knives, cutting up cardboard and making some delightful diversion for his children or grandchildren. He once built a whole city with my children, a settlement, he exclaimed with great glee, that they had "hacked out of the wilderness." On another occasion, he sliced his thigh open with one of those knives, while making a cardboard space helmet, complete with air tanks, for my brother Glenn. Mom had just learned to drive, but the emergency required her to get behind the wheel and rush him to a doctor. He was left with a noticeable scar, about the same size as the scar on my shoulder. In a way the two wounds united us.

Unfortunately, we were not united by his love for tools or his skill at using them. I was always inept and impatient with any implement other than a typewriter, and to this day, three simple words can strike fear in my heart: "Some assembly required." When homeownership or grandparenthood demand my attention to a deconstructed bicycle or bookcase, and I'm trying to join two balky pieces together that don't quite fit, I can still hear Dad's stern admonition in my ears: "Don't force it!" In addition to his precious X-Acto knives, Dad also collected rulers, particularly tape measures and white wooden carpenter's rulers that folded neatly into a pocket-sized package. Many of them were marketed as "unbreakable," but Dad knew the temptations of childhood well. "I know it says unbreakable," he would tell us constantly, "but if you try really hard, it will break. So don't try." It's funny. A man who never picked up a shirt or washed a dish was meticulous when it came to his tools. They were always hung up, often on pegboards, and today, eight years after his death, they remain undisturbed on the walls of his garage. Smaller items, like washers or screws, were always stored carefully in cardboard cigar boxes. Every time I see a box like that I think of Dad and the lessons he taught me. The ones that took, and the ones that didn't.

In our baby book, Mom mentions that Dad "has designed a toy called 'Constructo' that we're all very proud of." It was a building set made of cardboard pieces, and while it was never marketed, it contained a basic insight that led to many other successful projects. (My daughter recently bought a set of cardboard blocks for the guests to play with at her twins' third birthday party. Dad would have been thrilled.) With the war on, materials like wood, metal, and plastic were not available for toys. But kids still had to play with something, and by the next year, 1944, Dad had produced his first book. It's called *Let's Find Skipper,* and there on the cover is my first byline, or half a byline, to be precise: "Story by Jeffrey Victor." The story is pretty simple: Judy and Mike want to have a picnic, but they can't find their dog, Skipper, a brown spaniel. So they send their animal friends—Rusty the Rabbit, Danny the Duck—out on a search mission. Nobody can find the wayward pet until Judy suggests they look in the doghouse.

And there is Skipper, fast asleep. More noticeable than the story is the design of the book. Each page is a different size and shape, featuring a cutout of a different animal. When the pages are all put together, they form a montage of Judy and Mike surrounded by their furry friends. This is Dad and his knife and his imagination at work. He realized that books could also be toys, that kids could have as much fun turning the pages as reading the story. The story does include one element Dad always despised—talking animals. For all his playfulness, he never liked or understood fantasy very well, and prided himself on giving kids facts and information. *Let's Find Skipper* was the first, and last, of his books to contain a character like Lulu the Lamb, who says, "I think Skipper is chasing woodchucks in the pasture."

Over the next year or two, Dad published three more books that refined this approach. He called them "build-up" books, and they all followed the same pattern. *Let's Have a Parade*, *Let's Have a Store*, and *Let's Have a Farm* each start with a boy and a girl and a project. The books are bound at the top, not the side, and the first page is pretty devoid of illustration. In *Store*, for example, Peggy and Dan are in an empty room, contemplating the shelves they'll need for canned peaches and peanut butter. Flip up the next page, and the shelves are now against the wall. Then they need a cash register and counter, and sure enough, those items appear on the next page. Seven more pages add tools, toys, and even watermelons, but nobody comes to the store. Finally, they stock double-sized ice cream cones and chocolate marshmallow fudge nut sundaes. And on the last page, the store is jammed with hungry kids.

The books were a success, and even earned on the front cover a "commended" seal from *Parents Magazine*. But Dad needed a new gimmick, and his next effort was a real breakthrough. It was called *The Shoe Book*, and contained a bunch of familiar nursery rhymes—The Old Woman Who Lived in a Shoe was the first—plus some new verses written by Margaret Glover Hansl. In this case the "design" is credited to Jeffrey Victor, not the story, and the design is what made it such a smash hit. The book was actually cut in the shape of a shoe, with four holes near the top, and came complete with a shoelace, so kids could practice tying their shoes, an important skill in the pre-Velcro era. The book sold more than a million copies and has been mimicked many times over.

My favorite, however, is the next one, published in 1946. It's called *The Train Book*, and it came directly from one of our typical outings. Dad used to push Marc and me in our stroller to a train station a few blocks from our house. As the trains went by, he'd explain what each car was for and how it worked. He was a gifted teacher, we loved his stories, and he realized the material was there for a book. The cover simply says "By Jeffrey Victor," and on this one we had actually earned our credit. A sample page explains the role of the engine: "The engine pulls the train. This is how it works: a hot fire burns in the engine. The fire makes steam and the steam pushes the pistons that make the wheels go round. The place where the engineer sits is called the cab. In the cab are all the handles and buttons and levers that blow the whistle and ring the bell and light the light in front and make the engine go fast or slow." I can hear Dad's voice as I read those words: clear, patient, not too complicated but not too simplified, either. "This is how it works" could have been the motto on the family crest, with a pen crossing a knife on a field of cardboard. An added bonus: The book was actually in the shape of a train, so kids could play with it as well as read it.

By this time the war was over, of course, but toymakers were still adjusting and the market was still open. Dad produced three more books in the next few years, and by then he was putting his own name on the cover as the publisher. After *The Train Book* came *The Fire Engine Book,* which added another new wrinkle: Not only was the book shaped like an engine, it had wheels on the bottom that actually turned. "I was the first publisher to put wheels on a book," he once boasted, "and I still see that copied, forty years later." His next effort was *The Fix-It Book,* stories about carpenters and mechanics and such. In this case the book was bound at the bottom, and the top was in the shape of a handle, so it could be carried around like a toolbox. As I write I'm looking at a copy of *The Fix-It Book,* published in 1949, and the opening illustration looks a whole lot like our neighborhood in Bayonne, with shops lining the street that fix clocks and locks, shoes and pants. This was not a disposable society. Our parents were shaped by the Depression, and they fixed everything before buying something new. I had relatives and friends who did the jobs and ran the shops depicted in *The*

Fix-It Book, and the watchmaker is named Emil, in honor of my great-uncle, Miriam's brother, who was indeed a watchmaker.

Eventually Dad lost the business. By the time I was old enough to care, none of Dad's books were left. Flooded basements took a toll, but my parents were curiously unsentimental about them. That part of their lives was past, and I never read any of Dad's books to my own children. But years later the Internet, and my sister Laura, came to the rescue. She scoured used-book sites and assembled a complete set. A few of the titles were widely available, and for Passover a few years ago, she presented each of her three brothers with a copy of one of Dad's books. It was the best present ever, I cried when I saw it, and I asked her if she would get me a complete set as well. She did, and as my children were thoughtful enough to produce children of their own, Laura started collections for the next generation. I actually have a picture of myself reading *The Shoe Book* to Regan Roberts, our first grandchild. She's not very interested, truth be told, but that doesn't dim the symbolism of the moment. When I hear that our next two grandbabies, Jack and Cal Hartman, love to pull books out of their bookcase to amuse themselves in the morning while their parents sleep, I know Dad would be pleased.

Nothing connects me more to my own past than reading and telling stories to my grandchildren. They call me "Teebs"—apparently kids can't say Steve very well, since Marc called me "Tee-Tee" as a child—and when we're together I always have two jobs: bath master and storyteller. For some wonderful reason, they all love tales about trains, particularly Cal and Jack. We read stories every day from a volume containing dozens of classics, but one of their favorites is *Freight Train* by Donald Crews, which they can now recite by heart. There's something about the sounds and rhythms of trains that kids find compelling. Cal approached me recently with two pieces of colored paper covered with scribbles. He instructed me to tape them together to make a book. After I complied, I asked him what the book was about. It's about trains, he said. I just sat there, seized by emotion. Almost sixty years after this tyke's great-grandfather had written *The Train Book*, he was, at age three, joining the family business. My granddaughter Regan prefers stories about air travel, and she has a playhouse with a toy phone and a steering wheel that she calls her "hairplane," and during a recent visit she asked if I would go on a flying trip with her. When I agreed, she told me to go pack. I started to fill a pretend suitcase but she insisted, "No, Teebs, you have to pack for real." She also told me what to include: a toothbrush, a bathing suit, and pajamas. As she explained patiently to her dim-witted grandfather, "This is going to be a very, very long trip." The "hairplane" took off and Regan announced that it was time to take a nap. She stripped completely, put on her fuzzy pink pajamas, and directed me to disrobe as well. I drew the line, settling for a pajama top over my shirt, and as we settled in for our nap, she asked me to tell her a story. The child is deeply devoted to the color pink, and I started weaving a tale about

Queen Pinky of Pinktown, where the main cash crop was pink lemonade that fell instead of rain. As I talked, I realized there was no place on earth I'd rather be at that moment, passing on a family tradition of words and writing to the next generation. My grandchildren, too, live in their fathers' houses.

THE STICKBALL CHRONICLES

The war was over, my uncles came home, and our neighbors returned to The Block, intent on resuming their lives, restoring normalcy, having more children, building their businesses, and securing their status in the solid, prosperous American middle class. Serving in the military and fighting for their country was a big step in the Americanization of my parents' generation, although their parents still talked in thick accents and read newspapers in Yiddish. Our family name was now Roberts, not Rogow, and I was ready to venture beyond The Block, to go to school and see a bit of the wider world. But I didn't go very far. No. 3 School (another imaginative name) was only two blocks away, and when we walked there as small children, my folks insisted we take an indirect route, so we could cross two big streets at cor-

A Rogow family portrait on the front porch of Abe and Miriam's house in about 1948. Front row from left: *My brothers Glenn and Marc, Grandma Miriam, me, Uncle Bussy.* Back row: *Dad, Mom, Bussy's wife Mickey, Uncle Murray.*

ners with traffic lights. The detour took us right past the house Dad was living in when he met Mom.

While the neighborhood was pretty homogeneous, I started learning some social and economic distinctions. The poorer Jews lived on streets in the Twenties, sometimes in apartment buildings. The middle group, including us, lived in the low Thirties, usually in two-family homes. North of Thirty-fifth Street, a few streets with real names had been carved out of a bay-front estate, next to the park. And on those streets lived the rich kids in one-family houses. For years I thought only wealthy people lived on streets with names—Benmore Terrace, Wesley Court—and the rest of us settled for blocks with mere numbers and letters. Obviously I had never heard of Fifth Avenue, and I was totally oblivious to the fact that in Abe's heyday, the Rogows had owned one of the more luxurious houses on Wesley Court. But that was not our territory anymore. The families on Wesley Court belonged to country clubs in the suburbs and played tennis and golf. On Thirty-first Street we played stickball in the street. I didn't start playing tennis until I was forty and Dad didn't play golf until he was almost seventy.

These distinctions were seared so deeply in my mind that I dreamed about them recently. In the first scene, I was talking to old friends from high school and saying, to no one in particular, that forty-five years ago I had been in classes with Barrie Posnak and Lois Turtletaub and others in the room. I say friends, but classmates would be more accurate, since I didn't have many friends, and certainly not girls like Barrie and Lois, who were part of the in-crowd at No. 3 School and usually lived in that fabled terrain of named streets. In the second phase of the dream, I'm walking back to Thirty-first Street and noticing how the houses have been freshly painted and new shops have been opened selling pastries and expensive coffee. I'm not sure what that meant. Perhaps I just want to be proud of the old neighborhood, assured that it's doing okay. In my most common dream about Bayonne, I'm simply back on The Block and saying, to anyone I meet, "I used to live here." I say it happily, even eagerly. This is my turf, my home base. I'm one of you.

I love the recruitment ads for New York City teachers that say: You remember the name of your fourth grade teacher, who will remember yours? I do in fact remember the name of my fourth grade teacher. It was Ida Zeik, and she was the first person outside my family to see some spark of ability in the buck-toothed, big-eared Roberts boy who wrote compositions for her about such imaginative subjects as the wonders of nature. By the sixth grade, I was running a little newspaper we put out in class, and from that point on, I thought of myself as a writer. The encouragement and approval I received at home certainly pushed me in that direction, but there was another reason I found that identity so inviting. Like the two shy letter writers who lived a block apart, and became my parents, I found it much easier to express myself on paper than in person. I could be someone different when I wrote, someone confident instead of confused, suave instead of sweaty. I couldn't talk to Barrie or Lois, but I could write things they would read.

Even though the neighborhood seemed sealed and secure from outside influences, the Cold War flickered at the edge of our awareness. We had bubble-gum cards documenting all the evils of the "Red Menace," with the Chinese soldiers in Korea invariably depicted as grinning, bloodthirsty butchers. We had air-raid drills at school, where we crouched under our desks or huddled in the basement, preparing for the day when Russian bombs rained down on our village. Those drills scared the hell out of me, and since we lived directly under the flight path to Newark Airport, I would lie awake at night, convinced that the planes I heard were Russian bombers. Finally, I devised a game to ease my terror. If I heard a plane, and counted to twenty, and no bombs fell, I could relax and try to sleep. Until the next plane. Perhaps my fears were not entirely imaginary. An antiaircraft battery, equipped with Nike missiles, was installed in the park only blocks from our house, the same park where Dad took us as small children to swing on the jungle gym and sled down Double Hill. I guess others were worried about the Russians, too.

The other outside influence that filtered into the neighborhood was television. It came to The Block in the early fifties, beaming Milton Berle and Howdy Doody and Captain Video into our cramped living rooms. In this age of cables and satellites and unlimited choices, it's hard to imagine a world where TV sets received only three or four channels and everyone watched the same shows. If you missed Uncle Miltie on Tuesday night, you were shut out of the conversations the next day because everyone was discussing his skits and repeating his punch lines. But we weren't allowed to watch much TV, a deprivation that I've always been grateful for. Dad saw TV as a huge threat to his book business, and of course he was right. And he hated shows like *The Life of Riley* with William Bendix, because he thought fathers were always depicted as buffoons. Right again. Above all, my parents were readers, and they saw television as a superficial medium that would waste their kids' time and distract them from more enriching pursuits. (Like stickball?) Years later my wife and I lived in Greece when our children were small, and since all the channels were in Greek, the kids had no access to television and spent a lot of time reading. They've always appreciated that side benefit of a childhood spent abroad, but when they came back to America at ages seven and nine, they were social outcasts until they figured out who the Brady Bunch was.

I understood about being a social outcast. I lived my entire childhood with my nose pressed against the window of the room where the in-crowd gathered. But one way I could connect with other kids was through trading cards. The neighborhood contained none of the coffee bars or pastry shops I dreamed about, but there were plenty of "candy stores," as we called them, strange little places that sold newspapers and cigarettes and packs of bubble gum with baseball cards included. (They also had banks of pinball machines in the back, which I never learned to play very well.) The gum tasted like the cards, so we threw it away, but that wasn't the point. The cards were the prize, with the player's picture on one side and his statistics on the other, and card-flipping was a contest that tested both your skill and your nerve. The first flipper would toss out, say, five cards. The trick was to get them all to land on the same side, making a match harder. If the second flipper did match the first, he won the cards; if he failed, the first flipper won. I was pretty good, but out of my league when I faced off with Richie Wyroska, who lived around the corner on Thirty-second Street and cleaned me out regularly. Occasionally companies would try to market football or basketball cards, but those professional leagues were still new and we didn't pay much attention. At one point I did buy some football cards, but the pack was empty.

Dad was outraged on my behalf and wrote to the company, which responded by sending me a whole set. It was nice of them to do that, but like Grandpa Harry bringing me a bagful of chestnuts, the gesture ruined the fun of collecting something on my own.

My main connection to kids my age was through sports. Even though the park was nearby, The Block was our ball field and the players all lived within shouting distance. Traffic was pretty sparse, except for once a day, quitting time for the workers at the doll factory on the edge of the bay. When the whistle blew, we had to interrupt our games as the parking lot emptied. (The factory and its lot are now gone, replaced by about forty new homes.) The earliest game we mastered was called homers. Most of the front stoops on The Block were edged by flat panels of concrete built on a slant. The "batter" threw a ball against the concrete slab, and as it bounced into the air, the "fielders" tried to catch it. If they did, you were out. If the ball landed safely on the other side of the street, you had a homer.

As we got older, homers gave way to a more complex game, stickball. Each block seemed to have a slightly different version, but on Thirty-first Street, home plate was a round sewer cover about six houses away from us. The pitcher threw the ball to the batter, who tried to hit it with a sawed-off broomstick. If your hit reached the next sewer cover on the fly, it was a home run. It's a little-known fact in sports history, but I actually broke Babe Ruth's home-run record of sixty in a season, by a wide margin. I also had an unfair advantage. Dad had bought us an old pool table, which he installed in our attic, and it came with a bunch of used pool cues. Those cues, tapered slightly and weighted at the thick end, made excellent stickball bats, far superior to the spindly versions my rivals were forced to use. But I was, after all, the grandson of Abe Rogow. He would have been proud of my ingenuity.

The Block was also a fine football field, with only an occasional Studebaker or Oldsmobile parked on the sidelines. As the seasons changed, the sewer covers turned from home plates and home-run markers to goal lines. When I was in the fifth grade, my folks bought me a real football uniform, and for some reason I decided to wear it one day for our game of two-hand touch. Marc was playing quarterback and threw a long pass down the left sideline that I snared in the end zone. The only problem was that a rather large Buick was playing defensive back for the other side. My shin hit its bumper and snapped in two. I held on to the ball, the touchdown counted. But when I couldn't get up, I knew I was in trouble. My playmates found a red wagon, loaded me into it, and called up to my mother, who was eight and a half months pregnant with my sister Laura. The poor woman let out a long sigh, got her keys, and drove me to the doctor. Just what she needed—a bedridden son to go along with the new baby. However, Mom was unflappable in moments of crisis. The inexperienced and anxious young mother was now a battle-hardened veteran of the parental wars.

I was still a shy and solitary child who didn't always work and play well with others. The other kids sensed my weakness and teased me with a determination only children can muster. A gang of boys once stole a hat I was wearing, and when they wouldn't give it back, I burst into tears. So I devised another pastime that I could play by myself. Dad had bought me a game called All-Star Baseball, which operated on a simple premise. Each major league player was represented by a round cardboard disk that had a hole in the middle and fit over a metal spinner. Numbers on the disk reflected a batter's real-life statistics, and where the spinner landed determined the play. I created a whole league of my own, complete with team records and individual statistics, and on many lunch hours, I would come home from school and play a game or two before returning to class. Nobody could steal my hat when I was playing All-Star Baseball. But in nice weather, there were softball games in the schoolyard at lunch hour, and I gradually gained enough confidence to join in. The left-field fence was a decent distance away, but the best hitters, guys like Mike Lipman and Roger Dreher, would regularly pummel shots into the street, even into Mike Tobin's yard on the other side. I could never do it. Not once. I still dream about going back to that schoolyard and belting one over that fence. I visited No. 3 School not long ago and the field is exactly the same as it was then, almost fifty years ago. But the kids who play there now probably wouldn't let me into their game. When you're 12, you think guys with gray hair and bad knees are about 106 and ready for the scrap heap.

Basketball has been called "the city game," and it's always

been at the center of Bayonne life. In that article for *The New York Times Magazine* on my high school reunion, here's what I wrote: "If Bayonne has a heartbeat, it is the thump of a basketball being dribbled on concrete by a twelve-year-old boy. Making the high school varsity is like being admitted to a pantheon of ineffable greatness. Your picture was enshrined on the dusty wall of a school corridor, your name chiseled into the base of a brass-plated trophy. You had torn a tiny shred of immortality from the sweatshirt of history." I never made the varsity and never tore off my shred of immortality, but I did spend many hours thumping a basketball against the concrete court in my backyard. Dad had rigged a basket on the garage, fashioning a backboard from an old wooden skid, composed of a half-dozen planks nailed together, once used to support the large stacks of paper he bought for his books. Rogows never waste a good piece of wood. He also erected an elaborate chicken-wire fence, designed to keep the ball from bouncing into the neighbor's yard, but it never worked very well. No matter. I loved the court and played in all kinds of weather. My hands would get slick and scaly with the cold. Sometimes I shoveled snow to clear the court. Usually, it was pitch dark before Mom poked her head out the kitchen window and demanded I come in. When I finally yielded to her entreaties, I'd find a way to keep practicing indoors, using a bent metal hanger wedged into the top of a closet door as a makeshift hoop. Often I played alone, but when other kids came over and joined the game, I was an all-star. I knew every crack in the concrete, every tilt of the rim. Talk about home-court advantage. The New York Knicks, playing before a screaming crowd in Madison Square Garden, don't have a larger edge.

My parents identified very strongly as Jews—remember Dad, seeking out a Jewish storeowner on his way west—but our household was totally devoid of religious observance. None of my male relatives—grandfathers, father, uncles—ever had a Bar Mitzvah, the coming-of-age ritual Jewish boys celebrate when they turn thirteen. But as we approached that landmark age, Marc and I asked our parents if we could have a Bar Mitzvah. My motives were hardly spiritual. I wanted the praise and presents showered on any celebrant. Moreover, I wanted to be like everybody else, and every other Jewish boy I knew was going through this ritual. I realize now, though, that another impulse was at work. Even at that age, I was already feeling a connection to my heritage, as a Jew and as a writer, and the two identities were always intertwined in my mind. I was the one, after all, who listened to Grandpa Abe's stories and remembered them. I was the one who later walked in his footsteps, on an ancient street in Tel Aviv and on a railroad platform in Bialystok. And even though Abe never prayed in a synagogue, he was still a Jew, a member of the tribe, and studying for a Bar Mitzvah was a way for me to join that tribe. And Dad, the man who had changed his name to disguise his Jewishness, wrote to me years later that because of our decision, the whole family "rediscovered our identity as Jews." While most of the congregations in town were Orthodox or Conservative, some young professional families had recently started a Reform temple, where many of the rituals were modernized and conducted in English, not Hebrew. Our folks joined this new temple, and we took up our studies. Unfortunately, Marc and I were probably not related to Shmuel and Alter Rogowsky, the brothers whose houses I'd visited in Poland, the ones who sang the Friday night prayers so beautifully. Between us, we couldn't sing a note, and our rendition of the prayers lent a whole new dimension to the word "monotonous." The only relief for the congregation was that we split the Bar Mitzvah assignments in half, so they had to suffer through only one service, not two.

As I got older, I left the protective shelter of The Block more often. Going to temple, on the bus, more than a dozen blocks from home, was one example. What really pushed me into the wider world, however, was sports. I played basketball and softball on No. 3 School teams, competing against kids from other parts of the city, and I can still remember my batting average from seventh grade. When I was eleven, I tried out for Little League baseball, a big deal in Bayonne. The Rosenthals from Maiden Form had donated land next to their factory, where a state-of-the-art stadium was built in honor of their late son, Louis, who had died as a teenager. If you hit a really long home run over the left-field fence, you could actually break a window in the factory. Needless to say, I never came close. But I did get lucky. I got drafted by the Elks Club team, which was well stocked with talented twelve-year-olds, particularly Tommy Kavula, a hard-throwing pitcher who actually played a year or two in the Yankee farm system, and his cousin Johnie, a third baseman. Their family owned a bar, way over on the east side of town, the Catholic side, and when I met them there once before a game, it marked my first trip, on my own, into that alien territory. I played first base, and my only real job was to catch the ball thrown by the infielders. I did that reasonably well, and with my very modest contribution, we won the city championship my first year. We received jackets for our victory, black I remember, with our names stitched on the front, and I thought I was pretty cool. Damn cool, actually.

The next year, on Opening Day, the Elks team, as defending champions, rode in open cars down Broadway to the Little League stadium. I was feeling pretty cocky, after my first and only ride in a victory parade, and I took a seat in the stands for the opening game. A kid named Jimmy Petronick was pitching, and really getting hammered. I kept joking with my teammates about how bad Jimmy was, and finally a woman sitting in front of me turned around in tears and pleaded, "Please be quiet, that's my son!" I should have listened, and learned some humility right there, but my time was coming.

The problem was that all the star twelve-year-olds who had really won the title for the Elks Club were now too old to play. The Kavulas, Eddie Hetherington, Billy Foley. We had no experienced pitchers left on the team, so I was tabbed to start the first game. I didn't have a clue about pitching, but somehow we won. Every ball the other team hit well, and there were many of them, went right to one of our fielders.

Every ball we hit found a gap. But that was it for the season. I was terrible and so were most of my teammates. I think we won one more game the whole year. I still wore my black championship jacket, but with not quite so much swagger. And my father insisted that getting slapped around, game after game, was good for my character. Things don't always come easily, he would say, you have to work for them. I thought, at times, that he took almost too much pleasure in my defeats, but the lesson stuck. You don't always get drafted by the Elks Club, with a roster of star twelve-year-olds to pull you through. Sometimes, you just have to take your lumps.

I picked up some more lumps the next year when I played in a basketball version of Little League and was assigned to the VFW team, Veterans of Foreign Wars. All those endless hours in the backyard paid off, I was named to the league all-star team, and we reached the state finals against a team from Atlantic City. Unfortunately, the game was played in Atlantic City, and the crowd, and the referees, all sided with our opponents. Talk about a home-court advantage. It was better than me knowing all the cracks and tilts in my backyard. We lost narrowly, and this time my father didn't chalk it up to a life lesson. He was furious. I still am. That's another dream that recurs occasionally, going back and winning that championship game.

Sports taught me to handle defeat, but they also taught me to appreciate the culture of my hometown. You can tell a lot about Bayonne from the names on the shirts in various sports leagues. I remember the feel of the raised letters on my jerseys that spelled out Elks and VFW. Other uniforms said Kiwanis and Rotary, Knights of Columbus and Unico, an Italian social club. These associations were the heart of civic life, and we were proud, even as kids, to wear their names and colors. Neighborhood bars and taverns played a big role as well. They were social centers, not just watering holes. Walking around a city like Washington, you see young people wearing T-shirts from their colleges—Syracuse or Stanford or Slippery Rock—while in Bayonne, a lot of folks who never went to college wore shirts emblazoned with the names of local gathering places like the Clear Deck or the Venice or the Amble Inn. They served the same purpose as those college shirts. They said: This is where I belong, these are my people.

TV brought more than Howdy Doody and Captain Video into our lives, it brought Joe DiMaggio and Yogi Berra and the rest of the great Yankee teams of the forties and fifties. But even before we had a TV, I had started listening to baseball on the radio, and my first year as a real fan was 1949, when I was six. I'm still a Yankee fan today. That's my team, those are my people. As a small child, I had an old radio with a cracked brown plastic case that had been glued back together many times, and often I had to hold it up at a certain angle to catch the play-by-play patter. But even through the static, there was no better way to spend a summer evening than listening to Yankee announcer Mel Allen, and yearning for word of another "Ballantine blast" by Joe D. or Yogi or Tommy Heinrich or even Tommy Byrne, that "gooood hittin' pitcher," as Mel, a Jew from Alabama with a slight Southern drawl, invariably called him. "Going, going, gone," Mel would exult, one of the sweetest sounds a six-year-old boy could ever hear coming out of a radio. Each of the three New York teams was sponsored by a different beer—Ballantine for the Yanks, Schaefer for the Dodgers, Knickerbocker for the Giants—and team loyalties carried over to beverage choices. For years there was a Ballantine brewery near Newark Airport, and while I never drank the stuff I liked seeing the brewery, it reminded me of Mel. To this day I have a button on my car radio tuned to the Yankees' station in New York, 880 AM, and on a clear night I can get the broadcast in Washington after about 9:30. At home I subscribe to an Internet service that gives me all 162 Yankee games on the radio for about $12. As my grandparents would have said: Such a bargain! I also have a satellite dish, and for a long time I thought the only TV signal it received was for Yankee games. Imagine my surprise when I discovered it got hundreds of other channels! I once interviewed former Senator Jack Danforth, later the ambassador to the United Nations, on the issue of sports franchises moving to new cities. That led to a discussion of hometown loyalties, and we discovered that we shared a common vice: We'd sneak out to our cars in the middle of Washington dinner parties to catch the ball scores. The only difference was that he rooted for the St. Louis Cardinals.

Dad casually followed baseball as a kid, and during the World Series, would join other fans gathered in the street outside the headquarters of the *Bayonne Times*. The paper had a large sign in the shape of a baseball diamond, and each play was updated as small figures were moved around to simulate the game action. Dad got more interested in the game when I did, and he rooted for the Dodgers because they had signed Jackie Robinson and broken the color barrier. I lorded it over him throughout the fifties, as my Yankees regularly dispatched his Dodgers in the World Series, but when the Bums left Brooklyn for the West Coast, he shifted his loyalties to the Mets. So in 1986, when the Mets won the World Series, I called him up and said, "Dad, here's my present to you. I'm going to sit on the phone as long as you like, and you can gloat about the Mets." It was a small enough gesture, given all those years he had suffered through my boasting about the Yankees.

One of the paper companies that supplied Dad's publishing business had a box in Yankee Stadium and he'd take Marc and me to a game once or twice a season. The letter Dad wrote to the *Crimson* in 1963 was partly about baseball, and he included this passage: "Steven and his brother . . . were about seven or eight years of age when I first took them to Yankee Stadium. It was an exciting afternoon to see Joe DiMaggio, Tommy Heinrich and other heroes of the bubblegum picture cards come to life. I have never since achieved the omniscience in the eyes of my children as when I explained why they should bring in a left-handed pitcher, or bunt, AND THEY DID! As I recall, the boys had hot dogs in the third inning, peanuts in the sixth, and ice cream in the ninth. (I had beer.) We went home laden with scorecards, pennants, ice cream stains on our shirts and sunburn on our faces."

Dad had never been to a postseason game, however, and in 1996, when the Yanks were playing Baltimore in the playoffs, my parents were coming through Washington on their way to Florida just in time for a game at Camden Yards. I got us tickets and we went together, and I found myself asking him all the questions he used to ask me long ago: Dad, are you hungry? Dad, are you cold? Dad, do you have to go to the bathroom? We were both keenly aware of the role reversal. The son was now taking care of the father. And since he died seven months later, it was the only postseason game he ever attended. To this day, whenever I go to Yankee Stadium, I look for the spot where we used to sit and wave a silent salute to Dad.

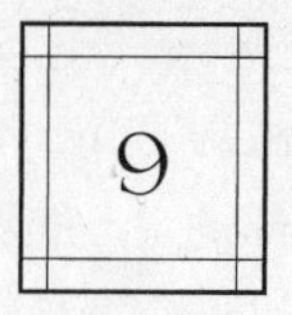

TOUGH TIMES

When Dad's books were selling well, my folks bought a modest house in the Jersey countryside near the town of Long Valley. The original structure dated to colonial times, with hand-cut pine floorboards and a bricked-up Dutch oven in the fireplace, and its charm compensated for its creaks and cracks. It was a place to get away—from The Block, from Bayonne, from a city house that was increasingly crowded with growing and demanding children. Over a period of six or seven years, we spent most of our summers in the country, and while there were no stickball games to occupy my time, I listened to a lot of Yankee games and honed my athletic skills by standing in the middle of the unpaved road next to the house and whacking stones with an old baseball bat. If I hit a ball, I'd have to go chase it, and that was too much work. I discovered fishing in those years, a perfect activity for a boy who didn't make friends well, and I spent countless hours on the banks of nearby ponds and streams, staring into the water and waiting—almost entirely without success—for a bite. However, one day my patience was rewarded. I was eight at the time, and fishing from a bridge down the road from our house, when I hooked a sizable trout. I was so surprised that I was barely able to wrestle the fish ashore and tote it safely home. It was a fall weekend, I had already started fourth grade, and consuming this symbol of my sporting prowess for dinner hardly satisfied my need for attention. We kept the fish on ice for the rest of the weekend, and on Monday morning I brought it proudly to class. My teacher was deeply impressed, first by the size of my catch, and then by its smell. She called Mom, who dutifully came to school and retrieved the trophy before its increasingly pungent aroma drove my classmates into the street.

Long Valley was not that far from Bayonne, but the east–west roads in New Jersey were poor, and the trip usually took several hours by car. Often we'd drive home on Sunday nights, with shows like *Our Miss Brooks*, later a TV series, playing on the radio. But it was dark, and a bit spooky, and we were plagued by a constant fear. Dad would run out of gas. No matter how many times Mom bugged him, he insisted on taking chances, as if driving with a full tank was only for wimps. Every warning or complaint was met with the same retort: "We have enough gas to get there and back." It became a family watchword, said with some bitterness, I might add, since we spent endless hours of my childhood stranded on the side of a dimly lit highway, while Dad foraged through the night, gas can in hand, looking for an open station. Most of the time, Dad knew he should listen to Mom's advice, but at times his stubbornness showed through, and the more she urged him to stop for gas, the more he resisted. At those moments he could be too much like his own father, ornery and obstinate.

Even when Dad was working in New York during the week, he'd come out to the country for weekends and vacations and take over the morning shift with us while Mom slept late. His favorite—and virtually only—culinary specialty was "eggs with the gas turned off." He would brown butter in a hot frying pan, then take it off the heat and let the scrambled eggs slowly congeal. Marc insists the whole thing was a mistake, that he once burned the butter and devised his recipe on the spot to cover up his error. No matter. I still remember the soft buttery taste of those eggs. One of Mom's specialties in the country was raspberry jam. A large berry patch grew virtually wild on our property, and several times a summer we'd put on long sleeves to ward off the thorns and pick the fruit. After making the jam, Mom would seal the jars with hot paraffin, so when you opened a fresh batch, you had to break through the waxy topping. Somehow it was sweet and tangy at the same time, the best jam I've ever tasted.

But no food in the country could rival the Cult of Corn. No Native American tribe, no Greek goddess of grain, worshipped this particular delicacy with the fervor of the Rogow/Roberts Clan. Dad would prowl through fields and farm stands, in search of the freshest, fullest ears, always pulling down part of the husk to double-check his choices. He would even call one neighborhood farmer to find out when the man was picking that day, so Dad could be there waiting when the harvest arrived. However, even a crop picked an hour earlier did not meet his exacting standards, so he started growing his own. Behind the house he planted a sizable vegetable patch, and corn was his specialty. When the harvest was ready, he'd start the water boiling even before he went out to the garden. After scrutinizing every stalk and weighing every option, he'd make his selections and then rush toward the house, shucking the ears as he went. Our job was to open the screen door for him as he bolted through the kitchen and plunged his prizes into the steaming cauldron. Mere seconds elapsed between picking and cooking. Eating took much longer, with most of the time spent commending the excellence of the fare we were privileged to enjoy. I'm a gardener myself now, and in our household the Cult of Corn has been replaced by the Temple of Tomato. But the rituals are similar, and I revel in the same sort of unbridled praise for my produce that Dad expected for his.

The country house came with a swimming pool, but not your normal, antiseptic, well-heated suburban version. This one was stream fed and icy cold, more lake than pool, and every spring Dad would dragoon some friend into spending a weekend, dredging a winter's worth of muck from the gray concrete floor. No shiny tiles or blue paint for us! (The system did produce one bonus: The muck was great for the garden.) We didn't have normal pool toys, either. Our rafts were wooden "skids," as they were called, which Dad got free from the paper companies he patronized—the same skids that also served as basketball backboards. I have two particularly vivid memories from that pool. In one, Dad is cradling me in his arms as I learn to swim. I felt scared but safe. He was there to catch me. In the other picture, years later, Dad is floating in the pool while his friend Harold Tucker is sitting on the edge and making notes. The two of them were stirring up a reform campaign for mayor of Bayonne, trying to elect a high school drafting teacher named Al Brady, and writing letters to the editor of the *Bayonne Times* signed "outraged veteran" or "perplexed housewife." It was an early insight into the deceptions of politics.

Since the house back in Bayonne was so crowded, it was hard to have company over, but in the country my parents could be more hospitable. One frequent guest was a newspaper writer named Bob Gardner, an old friend of Dad's who was part of his regular card-playing gang. They favored a game called pinochle, and the way they pronounced it, "pee-knuckle," I thought for years that it somehow involved urinating on your fingers. My cousins Susan and Carol were special favorites, since they were a few years older and starting to fill out their bathing suits in delicious and disturbing ways. Mom's old pal Tootsie Eisenberg and her husband, Sonny, were also regular visitors, but they kept kosher and brought their own food. I'm not sure if they were ever inducted into the Cult of Corn. Grandpa Harry came often as well, and Marc remembers a time when Harry taught him to drive a nail. They were working on the house, perhaps putting up new siding, and Marc's inept workmanship was leaving "owl eyes," as Harry called them, round hammer marks on the wood. So he showed Marc how to do it: one or two taps to align the nail properly, and then one swift sure blow, driving the shaft straight into the wood. "It was magic," Marc remembers, "it was completely impossible what he did."

My parents bought the house because it was affordable and convenient, but they were thrilled to discover that many of their neighbors were artists and intellectuals, seeking a refuge from New York City summers. Jimmy Ernst was a painter, Florence Lessing a dancer. Williams Phillips edited the *Partisan Review*, Grace Borgenicht ran an art gallery. Mom and Dad would never have known these people in Manhattan—after all, they spent many Saturday nights ordering pastrami sandwiches from Botwinick's on Avenue C. In Long Valley, accidentally and briefly, they mingled with the sort of people that Uncle Bussy, by now a successful writer in New York, saw every night. "They were very hip," Mom recalls. "It was a different lifestyle than we were used to in Bayonne." To say the least. One day Marc and Dad arrived at the Ernsts unannounced and apparently interrupted a nude swimming party. Such an event would have happened in Bayonne as often as the city elected a Protestant mayor. That is, never. One night was particularly memorable, December 4, 1953. Florence Lessing was appearing on Broadway in a new musical, *Kismet*, and she invited the whole Long Valley crowd, including Mom and Dad, to the opening. They were very excited, it was probably the only opening night they ever attended. But I didn't know, until many years later, that Bussy was there, too, reviewing the play for the *Hollywood Reporter*. I have his review, and in a paragraph listing "the dancing stars of the evening," the second name is Florence Lessing.

In 1949, something happened in the country house that would permanently change our lives. In my memory I was playing on the floor with my brother Marc and he couldn't stand up. In his memory he awoke in the middle of the night with terrible pains in his legs. My folks rushed him to the doctor and the diagnosis was crushing. Polio. He was placed in the isolation ward of a hospital and kept away from us for weeks. Mom came back to the car after leaving Marc and said to Dad, "You give them a child and they give you his clothes." There was only one way we could see him: Nurses held him up to the window, as we stood in the street, and we waved to each other. Marc felt "very lonely and frightened," particularly the day his mattress kept slipping. He rang the call button but no one came. When he tried to fix the problem, he fell out of bed: "I lay on the floor for a long while screaming."

The illness left Marc with a damaged leg, and Dad with a damaged conscience. He convinced himself that he was at fault, and over the years offered several different explanations for how Marc contracted the virus: He was playing with a stray kitten, he drank from a public water fountain, he used a public urinal. The day before Marc's symptoms surfaced, we were playing with our neighbors in the country, Betty and Bob Claus, jumping over a series of low fences they used to train beagles. Dad always wondered whether that vigorous exercise somehow made Marc's case worse. There was little scientific or medical basis for any of his theories, but that didn't matter. He couldn't forgive himself. I didn't know it at the time, but he started drinking and taking antidepressants, an unhealthy combination. Finally, it was Abe who sat him down and shook him up. When we came to America, the father told the son, we had no family here, no friends, no one to turn to. You are blessed, you are surrounded by people who care about you. And be grateful your child is alive.

Dad pulled out of his tailspin, but for both of my parents, the pain of Marc's illness never really went away, and a permanent cloud seemed to hover over the country house. "Could you put yourself in that position?" Mom asked me. "Could you imagine how you'd react? It was devastating. We had a beautiful, healthy child—I've never liked to use the word—who turned out to be crippled. It's hard to take, and at his age, the child was so defenseless." All true, and understandable. But Marc and I agree that Dad's reaction flowed from an even deeper well of feeling. The frightened boy who whistled in the dark to hide his fears, the young man who married the only woman he ever dated, saw a male child scarred and scared. His own anxieties, long suppressed, stormed to the surface. Dad looked at his son and wondered: Would he ever find a woman who could love him? Would Marc's psyche be stunted along with his leg? Years later, as a teenager, in a great display of courage, Marc ignored his disability and danced on the TV show *American Bandstand.* Dad could not watch the show, or even talk about it, without weeping.

Once Marc recovered, my parents faced a hard decision: How to treat him? And having a healthy twin brother, living in the same room, made matters more difficult. "I tried to be as rational as I could under the circumstances," Mom recalls. "I felt that if we treated Marc differently, if we indulged him or catered to him in any way, that would make him feel different. It would not be doing him a favor. So I decided that I would try to treat him as normally as possible under the circumstances, and for what it's worth, I think it had a beneficial effect." I agree. Marc emerged a strong, confident personality, a gifted teacher who's spent his whole career at Harvard, and now flies around the world to help developing nations organize their health delivery systems. But the process of instilling normalcy—of creating equality between the twins—was not always smooth.

At times, our parents wanted Marc to be like me. So they forced him to walk to school instead of driving him. At other times, they wanted me to be like Marc, and the symbol of that was a bicycle. Since Marc couldn't ride one safely, I couldn't have one, either. I felt aggrieved at the time, but in truth the lack of a bicycle didn't mean much in Bayonne. You could walk many places and take the bus to the others. In my entire life, I've never owned a bicycle, and I confess to having a phobia about them. They always seem unsafe and unsteady and I never feel comfortable riding one. Ironically, Marc came to like that form of transportation, taking his daughter to day care and commuting to work by bike, until two serious accidents dimmed his enthusiasm for the activity.

Even if our parents treated us the same, we were clearly different. I played sports incessantly. Marc joined our games occasionally—he threw the pass that led to my broken leg—but he spent even more time by himself than I did, reading books and making model airplanes in an attic hideaway. Psychologist Nancy Segal, who examined our relationship in her study of twins, *Intertwined Lives*, wrote: "Marc was a university scholar when he was eight. Will Roberts watched his young son 'delivering lectures' on the state of the world to anyone who would listen." Occasionally Marc's professorial manner got to me, and in a spasm of resentment, I smashed one of the models he had spent so much time constructing. Dad came up with an ingenious punishment: I was forced to make a model myself, so I could understand the damage I'd inflicted. It worked. I never touched his models again.

While Marc defied the doctors and recovered his ability to walk, he was left with a severe limp. In a curious way, however, his illness led us into joining the Cub Scouts. He loved the uniform, and he harbored a fantasy that if he joined the Scouts, he could somehow manage to march. The pack at the Jewish Community Center was in shambles, but our parents came to the rescue. Dad became the pack leader and Mom a den mother, and that experience symbolized much of my childhood. My folks were always there to lend me a hand,

and Marc was always there to give me a push. To the outside world we were always "the twins," or "the Roberts boys." We were always being compared to each other, and while that sense of competition made me frustrated and angry at times, it also drove me to match Marc's accomplishments. After Cub Scouts we joined the Boy Scouts, and Marc points out with glee that he made Eagle Scout six months before I did. But I never would have made it at all without him there, egging me on, and one moment symbolizes that for me. Bayonne had a Boy Scout camp in rural Jersey, not far from Long Valley, endowed by the Rosenthals from Maiden Form, the same public-spirited family who built the Little League stadium. Marc and I spent parts of three summers there, and one of the merit badges I coveted was hiking, a badge that Marc could never attain. One day a senior leader assembled a group of us for a twenty-mile hike. One by one the others dropped out, until I was the only one left. But I was on another planet. I was going to keep going, no matter what. It was after nightfall when we returned to camp, and I knew I had proven something important to myself.

Boy Scouts really worked for us. We learned self-sufficiency and responsibility. We wanted to win the clean bunk award, and worked hard at it. We battled fiercely against our rival campsites in weekly swimming races and scavenger hunts. On one notable occasion, when a hurricane blew through the camp, Marc and I decided to stay in our tent all night, while virtually every other camper fled to the safety of the mess hall. We drove our tent stakes deep into the ground, plugged a small leak with a sock, and stuck it out. When we emerged the next morning, safe and relatively dry, we had survived some rite of passage as meaningful as any Bar Mitzvah. Just without the presents.

Religious and ethnic identity touched everything in Bayonne, a place where the Polish-American Home was one of the biggest buildings in town. The two Boy Scout troops sponsored by Jewish organizations were always placed in the same campsite, with no non-Jews in the area. And since the Rosenthals were an observant Jewish family, they had built a kosher kitchen, and almost all of the Jewish Scouts ate there. But Marc and I, joined by one other pal, rebelled. We didn't keep kosher at home and saw no point in doing so at camp. Some of the other Jews started razzing us as *treif*, a Hebrew word meaning unclean, or nonkosher. The leader of our tormentors was a fat bully named Ira, and one day I just snapped. This time, I wasn't going to let anybody steal my hat. I got in a bloody fistfight with Ira that left me exhausted and exhilarated. And he didn't tease us again.

Marc's illness coincided with another serious blow, the end of Dad's book business. Although the two events don't seem connected on the surface, Dad's dark mood over his personal problems could have easily contributed to his professional decline. But the immediate cause was the rebounding toy market. As Mom recalls: "The whole thing collapsed when the war ended. All the old metal manufacturers got back into business," making toys widely available again. Dad got squeezed in another way as well. Simon and Schuster, a major New York publisher, had started Little Golden Books in 1942 to sell cheap, durable books for kids, and eventually Dad simply couldn't compete with such a big house and all its resources. He was basically living from book to book, and two ventures finally sank him. One, *My Flower Garden Book*, came with seed packets to plant your own garden. The way I heard the story, he got taken by his supplier, a true *gonif* (as Abe would have called him), an expressive Yiddish word meaning thief. The seeds never sprouted and neither did sales. The other disaster was called *A Child's Story of the Nativity*, and why my father decided to publish that title I'll never know. At least he knew something about trains. There was always the half-spoken belief rattling around in the family that somehow, Dad was being punished for his chutzpah. Why would a nice Jewish guy from Bayonne publish a book about Christianity? All I know is, we had hundreds of copies of both of them—the garden book and the nativity story—moldering in our garage for years.

Dad went to work for another company, Samuel Gabriel, and produced an outstanding series for young readers called Trumpet books. They had soft covers, and the same dimensions as comic books, but they delivered facts, not fantasies, on subjects like baseball. They were successful enough to merit a story in *Reader's Digest*, but two things happened to doom Dad's career. Gabriel's parent company got sold and the new owners had little interest in the children's book line. And Dad just never liked working for others. He wanted to set his own hours, sleeping late and working late, and he never fit comfortably into a corporate culture. Toward the end of his tenure at Gabriel, he was working alone, deep into the evening, when his pipe ashes started a fire in a trashcan. His bosses were not pleased. And when he went looking for another job, his reputation followed him. He was negotiating for a top editorial slot with a large New York publisher, when the subject of work hours came up. Dad allowed as how he wanted to set his own schedule. The publisher balked at the idea. Negotiations broke down. He didn't get the job. And he never worked in publishing again.

Tough times followed, and it must have been a nightmare for my parents, reliving the uncertainties of the Depression and seeing their hard-won security dissolve like packets of Jell-O in the sink. Dad tried different things. He worked briefly for the city of Bayonne, and started an auction market near our house in the country, an unhappy and unsuccessful echo of his huckstering days in Reno and First Street. One scheme involved selling paint-by-number kits through mail orders. The kits had to be assembled by hand, and when we were twelve or so, we got roped in as unwilling workers. You started with a square piece of cardboard, about fifteen inches on each side, which had perhaps twenty numbered holes punched in it. The holes were the size of small pots of paint, and our job was to put the right pots in the right holes. Green in the number 3 hole, yellow in the 8 hole, and so on. The work was boring and humiliating and I hated it. Instead of sympathizing with Dad's financial straits, and pitching in like a loyal son, I was frightened and embarrassed by his troubles.

Marc and I both remember going to school with patches on our pants and holes in our shoes, and when I needed something new, like a pair of sneakers, I couldn't bring myself to tell my parents. I didn't want to add to their troubles.

We survived this period because Mom simply would not let the family falter. She had decided long ago that being a traditional wife and mother was her chosen role in life, and she poured all of her talent and determination into that task. She had known adversity before. She had dropped out of college when her father's finances fell. She had warned her future husband, when Dad's Reno scheme was collapsing, that "we've got to stop being such goddamn fools." Now she was called on again to pull everything together, and she did. She showed no fear, at least to us. She was there, every day, an unblinking light of reassurance. She once said, "My strength is my strength," and that sums up her character perfectly. To be sure, her single-mindedness had a downside. Her fierce focus on her family could limit her larger view. Her kids used to joke that if you told Mom you were going to a peace march or a rock concert, her response would be: It will probably rain and the bus won't have a place to park. Once, when Cokie was newly pregnant with my parents' first grandchild, we were visiting them in New Jersey and decided to take a walk after dinner. After fifteen minutes or so, I noticed a pair of headlights, creeping along the street behind us. I turned to Cokie and said, "That's Mom." My wife wouldn't believe me, but of course it was Mom. Her light never went out, even when you wanted it to.

At one point Dad even went to work for his father, who now owned a small mobile home park in Bayonne. It was a lousy idea, given their long and tangled history, and Mom remembers it as a low point in their fifty-seven years of marriage. With an anguish that has not mellowed over a half century, Mom recalls that Abe went around town saying that his son was now driving his garbage truck. Eventually, the experience produced a positive outcome. Dad learned the mobile home business, wound up building and owning several parks of his own, and created a very comfortable living for his family. However, that was in the future. As the "twinnies" finished eighth grade and got ready for high school, something had to change. My sister Laura had been born three years before, and our flat now contained five bedrooms, a living room, a kitchen, and one small bathroom. That was it. All on one floor. Privacy was impossible. So the houses on Thirty-first Street and in Long Valley were both sold, and we bought a larger house on Thirty-third Street, between Avenue C and Broadway. We were ready for high school. And for the first time in more than thirty years, no one named Schanbam was living on The Block.

ABE AND BUSSY

After World War II ended, Abe found himself owning a large number of excess boards he had liberated from the Navy. All the chicken coops in Lakewood couldn't use them up, so he bought some property overlooking Newark Bay, on Lincoln Parkway, just three blocks from Thirty-first Street, and used the boards to construct a house. (Lincoln Parkway was an anomaly, a named street plunked down in the middle of the West Thirties, and there was nothing parklike about it.) Abe was no Harry, who could drive a nail with a single blow, and the workmanship—as well as the materials—left a lot to be desired. So did the building lot, which was too narrow and left little room on the bay side. No problem. Abe just filled in a chunk of the bay with truckloads of rubble to make himself a yard. He didn't own the land under the water, of course, but my grandfather didn't always trouble himself with formalities like property deeds. Even with the extra land attached, however, Abe couldn't sell the house, so he and Miriam moved in. But he hated the fact that neighbors had grown used to parking in front. So he got a piece of lumber and a bucket of yellow paint, and using the board as a straight edge, painted yellow lines in the street and the large legend NO PARKING. It worked. No interlopers parked there anymore.

We were regular Sunday visitors to the Rogow household, but Abe made few concessions to the presence of his grandchildren. Silence was required while he watched his favorite TV programs, and Glenn remembers thinking: "Why do they want us over here if we're not allowed to do anything? To this day I don't know the answer to that question." Actually, the answer was obvious. Abe considered himself a student of politics, and we were interrupting his studies. He always believed he had something to say—many things, actually—that the world should hear. To that end he privately published a small pamphlet called "A Plan for Immediate and Lasting Peace." I have a copy, badly damaged by too many flooded basements, but the author's name on the cover is clearly legible: By Abe Rogow, Member of Local 383, Carpenters' Union, Bayonne, N.J. I can't read most of it but here's the Foreword: "Many brows will probably be raised when the flyleaf, 'By a Carpenter,' is read. The first reaction will be, 'Why doesn't he stick to his square and saw?' Usually subjects like this are discussed by economists and sociologists. What this sick world needs is a fundamental reconstruction and replanning. Why couldn't a carpenter contribute some plan that would be logical? After all, a carpenter is trained in planning." This is pure Grandpa Abe. He didn't think small. Not only did he have a plan for peace, but an "immediate and lasting" peace. And if the *New York Times* and *Meet the Press* were going to ignore him, well, he'd just publish his proposal himself.

In Abe's view one thing this "sick world" needed was an end to the British Empire. This is a man who had resented and ridiculed the British land developers he'd met in Palestine forty years before, and Abe nursed his grudges. So in 1947, he telegraphed Winston Churchill, the British prime minister, challenging him to a debate on the subject of British imperialism. Abe heard nothing from Sir Winston, so he sent another wire to the PM's son Randolph, then traveling in America on a lecture tour. Apparently Randolph had met a plumber in a Denver hotel and wound up in a heated exchange. According to news reports of the day, Abe's message said: "I see by the papers where you debated with a plumber on the merits of plumbing. I am willing to donate a week's wages for any charity you designate, even if it is bundles for members of the English colonial and foreign office, for the privilege of debating you on the merits of the British empire." Randolph wired back accepting the offer, and plans were made to stage the encounter in a school auditorium, the biggest in town. The *Bayonne Times* ran a story headlined, CHURCHILL'S SON, LOCAL CARPENTER TO SWAP PUNCHES. Randolph insisted: "I was caught off guard by the plumber, you know. But I shall be ready for the carpenter." Indeed, he added, "I have wired Mr. Rogow to get in touch with my business manager and work out the arrangements." The debate was never held, apparently because the business manager blocked it. But on those Sunday afternoons, Abe was in training. I think it was Grandma Miriam who wanted us around and kept issuing invitations. But Abe couldn't be disturbed. Who knew when Charles DeGaulle might be available for a chat?

Abe fancied himself an expert on many subjects, and politics was only one of them. Sex was another. He once confronted his grandchildren with this question: What's the most important development of the twentieth century? Airplanes? Television? Genocide? Nah. Nah. Nah. He rejected all of our suggestions. Okay, Pop, we finally asked, what is it? His answer: the revelation of the female figure. When he was a young man in Russia, the most you could hope for was a glimpse of an ankle. Now, look what you can see on any city street. Let alone the beach! When our son Lee was born, he called in great agitation. Cokie was somewhat taken aback when her husband's grandfather told her: Don't circumcise the baby. Too late, Pop, she replied. But she was deeply puzzled. After all, circumcision is considered a sign of the Jews' covenant with God. I hear, Abe explained, that you have better sex if you're not circumcised.

Lincoln Parkway had no toys for us to play with, so while *Omnibus* or *Meet the Press* droned on, we spent endless hours stacking and restacking Miriam's poker chips, a kind of a do-it-yourself, early version of Lego. I would bring a ball along and practice tossing it into a Chinese vase that stood in the hallway. Or we'd explore the quirks and curiosities of the house: a closet that opened on two sides, in two different rooms; a primitive steam bath in the basement; tools that Abe painted a bright yellow to discourage thieves; and large stores of extra supplies. One of his brainstorms was to employ small sponges instead of toilet paper. He thought they could be reused; the rest of us thought they were disgusting. My cousin Pam once brought a boyfriend to meet her grandparents, and they received a lecture from Abe about his sponges. She never did that again. Perhaps the nastiest word in Rogow family lingo was "retail," and Abe would buy all sorts of things in bulk, even when they didn't make sense. On one of Mom's first visits to her future in-laws, the house was festooned with signs that read: EAT MORE PEARS. Abe had gotten a great deal on a shipment of fruit, but they were rotting too rapidly. His trailer park had a row of public showers to accommodate travelers, and he once drove all the way to Brooklyn to buy a gross of shower curtains wholesale. That's 144. You can live a very long life, and keep very clean, and never come close to using up 144 shower curtains.

Abe loved gadgets and was constantly buying them, but he could seldom get them to work right. On one of our Sunday visits, he insisted we listen to his new, stereophonic radio. He couldn't get it to tune properly, so he turned on his old radio, a big, stand-up Philco, and set it to one of his favorite stations. Then he fiddled with the new radio, trying to get it to play the same station. Or so he thought. When all was finally ready, we were summoned to hear this magnificent new sound, and positioned ourselves around the living room to absorb the full stereophonic effect. We oohed and aahed at the musical selection, and when it was over, Abe tried to shut off the new radio, but he couldn't do it. Even after he unplugged it, it kept playing. Finally, he realized the grim truth. The new radio had never been turned on. All we'd been hearing, and praising, was that old Philco.

The food at Sunday dinner was certainly plentiful, but not very palatable. Miriam was not what you would call an overly domestic person. Marc says the chicken was boiled into a bland mush. Glenn once asked for a hamburger instead of the main dish. I was still hungry after one meal and made myself a peanut butter sandwich. The family never let me forget it, and on every visit after that, I was expected to repeat my performance. I didn't really like peanut butter, but the ritual had to be observed. I don't think I've eaten a peanut butter sandwich since. To be fair, our view of Miriam's culinary talents was not widely shared. Her children swore that she was a good cook. My uncle Murray complained she was too good, always stuffing him with food instead of affection. Aunt Rose used her mother's pot roast as bait in snaring her husband, Hy. Years later, when I wrote a story for the *New York Times* about traveling in Israel, I wrote this line: "The food in Israel is just like my grandmother used to make. Unfortunately." I was pilloried by letter writers, who accused me of dishonoring my grandmother and disdaining my heritage.

They were wrong. I loved Miriam and I loved my heritage. I just wasn't crazy about her cooking.

The house on Lincoln Parkway actually led Abe, and eventually Dad, into the mobile home business. Abe's upstairs tenant, a man named Tony Perozzi, was part owner of a local saloon, The Clear Deck. Business was shaky, since all his friends and relatives came to eat and drink on the house. Tony needed more income, and he knew someone who was doing quite well running a small trailer park in Bayonne. Since there was some extra land behind the tavern, Tony hired Abe to hook up sewer and electric lines so he could rent out spaces for trailers. When the spaces filled up immediately, Abe decided this was a good business. He got some backing from his old friend Jake, who owned a thriving cafeteria in the Bronx, right near Yankee Stadium, bought some land from the city on the shore of Newark Bay, and opened Sunset Trailer Park.

But the land could hold only a few trailers and Abe had bigger ambitions. So he started filling in the bay, sinking barges and bringing in trucks from construction sites to dump their loads of concrete and brick. He'd done the same thing on Lincoln Parkway, but that was small potatoes compared to this project. And besides, he was now renting out the newly filled land and making a profit on it. The city cracked down, warning him to stop. So he put a sign up on a chain across the street that said in big letters, NO DUMPING. When the inspectors came by, he professed his innocence. I put a sign up, they don't listen to me, there's nothing I can do. Of course, when the trucks did come, he would simply take down the chain and show them where to leave their loads. During these deliveries Abe put on a priceless performance. He'd yell at the truck drivers, who were understandably cautious, being as they were about to fall into the bay: C'mon back! C'mon back! C'mon back! Then, just as they reached the edge, he'd scream: STOP! Eventually, I think, Abe simply paid off the inspectors, a time-honored Bayonne tradition. The sunken barges, which formed the foundation for the new land, would occasionally give way, causing a large sinkhole to appear in the front yard of some unsuspecting trailer owner. I visited Sunset recently, stood on the water's edge, and felt like yelling at the tenants: Be careful! I know what's under this land!

In Abe's world he was always right and everyone else was always wrong. He'd park his car behind his office at the entrance to Sunset. One day my brother Glenn watched as Abe backed his car out of his spot, careered completely across the street, and smashed into another car parked at the curb. No matter that cars were always parked there. Abe leaped from his own vehicle, screaming, "What sonofabitch parked that car there?" Abe was always playing the odds and the angles. He had stationery printed up that said Atlas Supply Company—there was no such thing—and when he needed a new tool, he'd write to the manufacturer, saying he was a wholesale distributor of wrenches or screwdrivers or whatever. He was thinking of handling the manufacturer's line, so could they send a sample for him to try? I'm not sure he ever paid for a tool in his life.

Abe was also bedeviled by enemies, real and imaginary. When stray cats started infesting the park, he bought a bow and arrow, convinced he could solve his problem through archery. I'm sure his weapon never posed a threat to any feline, and eventually he gave it to Glenn. He was also convinced that people were stealing from him, and there he had a point. They were. He built an elaborate cage in his office to secure his money supply overnight, complete with heavy bolts and bars of all kinds. But it was all a ruse. That's where the thieves would look first. The money was actually stored underneath a trash can in the toolshed. Then there was the check issue. Glenn worked for him several summers, and one of his jobs was to pay the bills. However, since every check had to be cosigned by Abe's partner, Jake, Abe would trek to the Bronx periodically and get Jake to sign a bunch of blank checks. Once Glenn made the mistake of using up the last signed check to pay a bill. At that point Abe sat his grandson down and imparted a life lesson: Never use up the last signed check. If Jake and I ever have a fight, Abe explained, I want to be able to clean out the account before he does.

For all his craziness, Abe could be very generous. He was constantly giving his grandchildren presents, like the suit he wanted to buy for me so I'd look sharp at the *Times*. But often his gifts came with the caution: You can have it if you don't ask where it came from. Some of his sources were, shall we say, less than legal. A lot of things fell off a lot of trucks in

Bayonne. At other times he could turn insensitive, even mean. After Dad went to work for him at Sunset, Abe actually fired his own son. It was an unforgivable act, and Mom never did forgive him. Glenn remembers the night it happened. Everyone in the family was crying. As a young man, Dad had considered his father a visionary, an empire builder, and he'd written to Mom saying that he wanted to be a hero in his father's eyes. Now here they were, twenty years later, with no empires in sight, fixing sewers and hauling trash, and I think they were both frustrated and disappointed. Dad hated driving Abe's garbage truck, but he needed the work. Abe resented Dad's need and didn't want him around. Each man reflected the other's failure, and it was just too painful. Abe was cruel to fire his son, but in the end Dad needed to break free, once and for all, from his father's grasp. And eventually he did.

I inherited Abe's interest in politics and ideas, his impulse to be heard by a wider world. But it was harder to be his son than his grandson. Toward the end of his life, Uncle Murray said with some anguish: "I grew up with Abe's belief that the world owed me a living." And Glenn believes Abe was a bad influence on our father: "Dad was contemptuous of people who put on a tie and went to work and Dad paid a price for that, in terms of many hard years. I think that contempt came from Abe, at least in part, and I'm not sure that was a good gift to give your child."

Abe's other son, Bussy, left his father's orbit and was better off for it. After returning from the war, he resumed his writing career, married my aunt Mickey—a witty and delightful woman—and eventually had two children, Maggie and Zack. His professional output was prodigious, almost as if he knew his time was limited. In addition to a full-time job at an ad agency, he was a ghostwriter for showman Billy Rose's Broadway column. In her biography of her brother, Polly Rose Gottlieb says Rose considered Bussy his "right arm" and "regarded him with deep respect." In his own name, Bussy reviewed plays for the *Hollywood Reporter,* and books and movies for the *Saturday Review*. His short stories appeared in *Esquire* and *Collier's*, *The New Yorker,* and *Redbook*. As his Navy writings indicated, his great gift was taking an ordinary situation and viewing it through a prism of wry humor. And he drew heavily on his family—his parents, his wife, his children—for his best material. One of his stories, "That Certain Flavor," was based on an old tale his mother used to tell. In Bussy's modern version, a woman named Polly Kingsley struggled mightily to please her persnickety husband, Larry. But no matter how long and hard she labored in the kitchen, he was never satisfied. Mother's cooking, he complained, "had a certain definite flavor that this lacks." Polly sank into a funk, until one day she met her friend Min, who enticed her into an afternoon cocktail, or two or three. She came home late, tossed a bowl of overcooked and underseasoned spaghetti at her husband, and waited for his reaction.

"Polly!" he said. "Polly Kingsley!"

"Anything wrong?" she snarled.

"Wrong?" shouted Larry. "I don't know what you've done, genius girl, but this is it."

"Hah?" said Polly.

"This spaghetti is just like Mother's!" said Larry. "That flavor, that certain definite flavor, it's all there!"

Bussy was entranced with his children and wrote about them constantly. In *Diapers for Maggie*, a high-powered executive spends a day taking care of his infant daughter. After running out of diapers—and buying a diaper service—he ends up using table napkins to keep her dry. In *The Great Grocery Cartel*, he describes life on Fire Island, a vacation retreat favored by New York writers. After his wife pays the exorbitant price of fifty-three cents for a can of tuna fish, he decides—in true Rogow tradition—to buy their provisions wholesale. "My own passion for the project was unbounded," he wrote. "When I was a boy my father had had a connection with a wholesale grocery firm, and I remembered with pleasure the excitement of delivery, the ripping open of cartons, the stacking of cans in cupboards, the sense of being well-supplied against flood, famine or Indian raids." One problem: The wholesaler would only accept orders in case lots. This dialogue follows between husband and wife:

" 'Why do we need a case of toothpicks?' she said, looking over my tentative order form. 'Do you realize that's almost five thousand toothpicks?' "

" 'We can use them for kindling, the children can play

with them, we can stick cocktail sausages on them,' I said cheerfully. 'That reminds me, we'd better order a case of cocktail sausages.'"

Needless to say, it all ended badly: "With a week of the summer to go every cupboard and closet in our cottage was still bursting with cans and cartons." Even a sign saying EAT MORE TUNA FISH would probably not have helped. But Abe's wackiness, spiced by Bussy's wit, made a tasty story.

In mid-1955, Lee was asked by the Army to plan a recruitment campaign. He kept turning down the assignment, but the Army persisted. I don't know anything about the Army, he said, I served in the Navy. If I do this, I need to visit some bases and observe their operations. Fine, said the Army, we'll arrange for a plane and a pilot to fly you around. On September 13, he took off from Mitchell Field on Long Island. The plane flew for ninety seconds and crashed. No one knows why. When Abe heard the news, he retreated to the attic on Lincoln Parkway and spoke to no one. Miriam murmured in Yiddish, over and over again, "My beautiful son." Dad was so devastated, a friend had to bring him home from New York. Forty-two years later, when Dad started to talk about Bussy at a family reunion, he began to weep and couldn't continue. Mickey never recovered, either, and once her children went off to college, she took her own life. Bussy's friend Herman Wouk wrote a letter to the *Saturday Review* that ended: "The snuffing out of his bright talent at thirty-six is a catastrophe for which there are no consoling words." He was right. There were no consoling words. But there would be other words, in other years. Bussy wouldn't write them. But at age twelve, I knew, somehow, that I would.

A year later, Miriam died of cancer. Abe was cut adrift. In the lonely house on Lincoln Parkway, he would write notes to himself in the dust that covered the tabletops. So that summer he invited Marc and me to take a car trip through New York State and into Canada. We made a bizarre traveling party, to say the least. The Everly Brothers were popular that year and we insisted on listening to their songs on the radio. Abe couldn't stand them. He always wanted to hear the news. We had constant fights over what to see and where to eat. When we stopped for the night, he'd take his ready cash from its hiding place—inside the spare tire in the trunk. When we reached our destination, Manitoulin Island in the middle of Lake Huron, he decided to give us a driving lesson. In truth, Abe was curious about whether Marc's disabled leg would permit him to drive. But he couldn't broach the subject with my parents, so he tested his theories when they weren't around. Marc, by the way, did fine. I drove off the road.

Once back in Bayonne, Abe decided to get married again, but rejected all Jewish contenders. They would spend too much time on their children, he felt, and not enough on him. Finally, he picked Pauline, one of Miriam's old card-playing "pigeons," a Polish widow who was as meticulous as Abe was sloppy. If he got up at night to use the bathroom, she'd make the bed again before he returned. It drove him nuts. Since he didn't want her prying into his finances, he kept his checkbook records in Yiddish, a language she didn't understand. A brilliant strategy, except his accountant didn't understand it either. The match didn't last long.

Abe had lost his son and his wife, but not his ego or his ambition. He read one day that private phone service was being established between the United States and Russia. So he decided to call his sister, Anna the Bolshevik, who had moved to Moscow, taught French to the children of Soviet commissars, and served as an official government translator. After several days the call went through. They'd been out of touch for many years, so Abe was very excited and full of questions. For instance: What happened to their brother Marcus, who had disappeared into the mists of the revolution? But Anna was curiously remote, responding with canned speeches about the glories of the Russian Motherland. Only later Abe found out that his call was one of the first to get through, and Anna was surrounded by government agents on her end of the line.

Not long after that, Abe went to see Anna in Moscow and was stunned by her poor living conditions. Even though she was a privileged old Bolshevik, she lived in a tiny apartment with her son, Spartak, a devout party member, and his wife. Before he left, Abe bought a refrigerator on the black market and gave it to his sister as a present. The son would have none of it, denouncing the gift as a product of capitalist exploitation. Then one night, as he was going to the bathroom, Abe saw Spartak in the kitchen, playing with the refrigerator and trying to figure out how the little light went on and off. The next morning the young man announced that they would, in fact, accept the gift, but only as a gesture of respect to their elderly relative.

Abe came home convinced that the gravest weakness in the Soviet system could be summed up in one word. Plumbing. He was so appalled at the sanitary facilities he saw in Russia, he was sure that if Soviet citizens had any idea of what bathrooms were like in the West, they would arise en masse and overthrow their rulers. So he launched a one-man crusade, sending copies of magazines like *Better Homes and Gardens* and *House Beautiful* back to his homeland. The whole caper contained echoes of an earlier time, when Abe had helped his sisters smuggle copies of Lenin's newspaper into czarist Russia. I doubt if the magazine missiles ever reached their targets, and I've long had a fantasy that somewhere in Moscow, an entire warehouse is devoted to crumbling copies of Abe's confiscated contraband. Of course, despite his nuttiness, Abe was right. The failure of the Soviet leadership to deliver consumer goods and decent living standards helped hasten their downfall.

In later years, after I joined the *Times*, he urged me to seek an assignment in Moscow. "The only thing, before you leave for Moscow, take a supply of toilet paper with you," Abe wrote. "The Russians in their hotels and homes use sliced Pravda for toilet paper." Then he was off on a riff about the Russian leader Nikita Khrushchev, who had stunned the world by pounding his shoe at the United Nations to make a point. Abe theorizes that Khrushchev left his shoe behind: "Well, Steve, I feel sorry for the poor slob. I feel like sending him a box of sliced Times paper, handkerchiefs, and a left shoe. I have a friend, a traveling shoe salesman, and he has carton's and carton's [sic] of left shoes. I wouldn't be surprised that the shoe was one of the causes of his downfall. With such a scarcity of shoes in Russia, the bum leaves a shoe with the imperialist." He points out that since his sister is so well connected with the party, I'd have special access to ordinary Russians and their private lives. "One thing you got to be carefull," he warns. "Don't walk to close to the buildings, or you will get hit by a brick. Not by a communist. No! Just they are laid loose, that is the bricks." Abe was laid pretty loose himself. He apologizes for not typing the letter, but he'd cut his finger slicing Polish kielbasa. He concludes: "Take Poland. There I go again. The first thing you know I will wind up sending American chop sui to Moe Tze-Thung. You got to be carefull with Moe. You know he has the 'thing.' The bomb. You cannot fool around with a fanatic that has a bomb. So that's about all. Best regards. Granpa AVREMEL."

I never got to Moscow, but the *Times* did send me to Athens, and on one holiday we took our kids to Israel. We saw the street that Abe had built in Tel Aviv and the night sky he had loved over the Galilee. Then we visited Jerusalem and stood before the Western Wall, one of the holiest sites in the ancient city. Abe always had a dream of returning to Israel for his final years, but when he eventually did move there, it was a disaster. He couldn't understand Hebrew and felt suffocated without his *New York Times*, so he left after a few months. But I could feel the spirit of the young pioneer who had once lived and worked in this land. At the Western Wall, the custom is to write a prayer on a piece of paper and leave it in one of the cracks in that ancient façade. I found myself writing out a prayer in the name of Abraham Rogowsky. Granpa AVREMEL.

NEWSPAPER BOY

We moved into our new house during the summer before high school, and it was a lot larger than the old one, a single-family structure with three floors and a sizable backyard. The third floor contained three rooms—a big front room for me and Marc, a smaller back room for Glenn, and our own bathroom. Mom would go weeks without venturing into our domain, which meant a higher standard of privacy and a lower standard of hygiene. But Thirty-third Street was too wide and busy for stickball games, so we had to settle for Wiffle ball in the backyard. Dad put a basketball hoop on the garage, and I soon received a memorable lesson in theology. Our new neighbors, the Eisenbergs, were deeply religious Jews and their son Barry, a year younger than me, was a devout basketball player. On Saturdays, the Jewish Sabbath, he would often come over and play in our yard. Here was the dilemma: Observant Jews are barred from doing any work on *shabbas*, and work includes carrying anything, even a basketball. Sometimes, Barry would leave his ball in our yard on Friday afternoon, but that was an unsatisfactory solution. He sought a Talmudic ruling from his grandfather, an Orthodox rabbi, and the compromise handed down was truly brilliant. Barry could bounce the ball over to our house, as long as he didn't carry it. Dribbling was play, not work, and thus permissible. I quickly learned the nuances of the new court—the hoop tilted downward to the left, making shots from that side easier—and even Barry's prayers and piety couldn't overcome my home-court advantage. Barry's grandfather might have been a rabbi, but mine was a Rogow.

We could walk to the high school, only about six blocks away, and on my first day I signed up for the student newspaper, the *Beacon*. I already knew what I wanted to be when I grew up. I now teach students at George Washington University who finish college without a clue about their future plans, as if graduation were some underhanded and unexpected trick played on them by the academic gods in the spring of their senior year. I never had their choices, but I never had their doubts and anxieties, either. I wanted to write, and my first assignment for the *Beacon* was the football team's opening game. We had a terrible team and the game was a blowout, so I looked for another angle. Along with my account of our devastating defeat, I handed in a column of notes and features about the team that I called Bee Buzzings, since our fearsome school mascot was, in fact, a bee. Both stories were printed in the first issue of the paper and I was on my way. I was still woefully inept when it came to girls, and the heavy metal braces I wore on my teeth, thanks to Uncle Bernie the Orthodonist, hardly helped my self-confidence. But on paper I was free from all my adolescent agony. Many years later, I saw a cartoon in *The Far Side* comic strip that summed up this feeling perfectly. A dog is pecking away at a computer and says, "On the Internet, they don't know you're a dog." In the *Beacon* they didn't know I was a nerd.

Since running the paper was not the coolest of school activities, I had little competition, so by sophomore year I was sports editor and by junior year, editor-in-chief. And that position gave me a shot at my first big journalistic assignment. A woman named Alexandra Zuck, who had preceded me at No. 3 School by a year, and once sang "Twinkle, Twinkle, Little Star" in a St. Patrick's Day program, returned to her hometown. She had changed her name to Sandra Dee, moved to Hollywood, and starred in two of the great teen movies of the late fifties, *Gidget* and *A Summer Place*. (She also kept a taste for local Polish food, and for years had care packages sent west from her favorite Bayonne deli.) The mayor proclaimed it Sandra Dee Day, and a flying squad of a dozen police motorcycles escorted her around the city. One event was a press conference with the editors of the three high school papers (the other two high schools were Catholic, Marist for boys and Holy Family for girls). I was stupefied. Here was Gidget. In person. A few feet away. My mouth hung open.

My questions died deep in my throat. Clearly, my interviewing technique needed a lot of work.

My first job for the *Bayonne Times*, the local daily, was delivering the paper, not writing for it. I did that one summer, borrowing a friend's bike, picking up the papers in early afternoon, and rushing to finish my route before the Yankee game started on TV. As children of the Depression, my parents were naturally thrifty—they remembered the days of cashing in soda bottles to pay for movie tickets—and I absorbed their habits. My income was, literally, a matter of nickels and dimes, and with an occasional quarter tip thrown in, I could clear $7 or $8 a week. I would dutifully take my earnings to a local savings bank, where I opened my own account and stacked my load of coins on the marble counter every week for deposit. But a year or two later, my duties and income were upgraded. The *Times* took one high school kid as a copy boy and the guy who was leaving the job, the older brother of a classmate, recommended me. I worked two hours after school every day and four on Saturday morning—fourteen hours for $14—and while I was already a newspaper boy in my own mind, I now became one for real. I was actually putting words in the paper, not tossing it on people's porches. Instead of lining up at the loading bay with the other delivery boys, I was working on the second floor, in the newsroom, and the feel and sound and odor of that place left an indelible mark. The newsroom was up a narrow set of stairs from the street and under modern environmental laws, might have qualified as a toxic waste dump. The pots of paste, used to slap pieces of copy together, oozed gobs of sticky goo down their sides. The pencils were thick, the paper thin, the coffee cold. Cigarette smoke filled the air and burn marks scarred the desks. The ancient typewriters coughed and wheezed and could barely catch their breath. The wire machines rattled with a jazzy, jagged rhythm that was both irritating and intoxicating. The sharp aroma of molten lead and moist ink from the pressroom on the ground floor never quite faded away. The assistant sports editor, whose specialty was bowling, kept the stub of a cigar jammed permanently into the side of his mouth.

Sitting at my typewriter in the newsroom of the Bayonne Times *in about 1959. It was my first real newspaper job, and I got paid $14 a week for fourteen hours of work. My duties included the movie timetable and the TV highlights.*

Another editor would regularly call out, "Take a stiff on seven." That would be an assignment to write an obituary. It was love at first sight, or smell, and I would spend a big chunk of the next thirty years hanging around newsrooms. When I joined the other *Times*, across the river in Manhattan, I loved staying late, until the first copies of the next morning's paper arrived in the newsroom. The ink was still wet, the paper still warm, and to a real newspaper boy, fresh-baked bread from your mother's oven could not have smelled any better.

However, even my work on the high school paper didn't prepare me very well for a real newsroom. When I got my first assignment—the exact subject eludes me now—I sat down and started writing it out in longhand. The editor just laughed and pointed to a typewriter. It took me several painful hours to finish. I had a lot of clerical duties—filling paste pots, changing typewriter ribbons, fetching fresh stacks of paper. The sports editor with the cigar, known universally as Kap, had me check the wire every day to see who won the daily double at whatever local racetrack was operating. He played the same numbers, 6 and 9, every day, and I had no idea of the sexual innuendo behind his wager. He never won. Another job was more fun. The comic strips would arrive from the syndicates in batches, weeks before they were scheduled to run, and I was assigned to assemble each day's installments in separate files. That meant I knew what was going to happen in the funny papers way ahead of everyone else, and I would drop tantalizing hints about the coming adventures of

Mark Trail and Prince Valiant and the other cartoon heroes. I also wrote the TV highlights box, which I cribbed directly from that week's issue of *TV Guide*, but I could insert a word or phrase of my own and practice my writing. That responsibility provided another important breakthrough. I'd buy the guide for fifty cents and then get repaid—my first expense account!

Occasionally, I would get to write an obituary, an important story for a local paper like the *Times*. And since Bayonne was such an immigrant community, and so many residents, unlike us, had not Americanized their names, spelling everything right was an acid test for any cub reporter. The Italian and Ukrainian and Lithuanian names were hard enough, but the hardest of all were the Polish names, all *cs* and *ys* and *zs*, and not a vowel in sight. But I quickly learned to turn those names to my advantage. I was assigned to write a column about high school news every Friday and payment—above and beyond my princely $1 an hour—was by the inch. I soon figured out that listing the football team's starting lineup, with all those long ethnic names, was worth about fifty cents alone. Even the basketball starters—Stasiulatis, Skolnick, Capitano, Osofsky, and Kurowski—netted me a quarter.

The paper's editor, a surprisingly quiet and gentle man named Bob Caldwell, would occasionally give me real stories to write. One of them, which got printed along with a picture, was about an immigrant painter who decorated the walls of the Police Athletic League building with murals. But Mr.

Caldwell's confidence in me had to survive a bad mistake. One of my regular duties was to assemble the movie timetable. The local theaters would call in the titles and the times, and I'd write them up for the next day's paper. But on New Year's Eve, one of the busiest days of the year, I put down that the DeWitt Theater was playing *Around the World in Eighty Days* at nine o'clock. Wrong. It started at eight. Half the audience came late, the theater lost a boatload of money, and my newspaper career almost came to a screeching halt. A hard but memorable way to learn the importance of accuracy, down to the smallest detail. My other sin stayed secret. I was playing in a lot of basketball leagues and became friendly with the guys who covered local teams on a part-time basis. They were always overworked and underpaid and happy to have my help. So I wound up writing stories about a number of games that I actually played in, without a byline, of course. And somehow I always managed to emerge a hero in the sports pages of the *Times*, even when I scored four points.

One day I was rummaging through the desk of a young reporter, looking for extra copy paper, when I came upon a manuscript. It was a long series he had written about Mafia influence in Bayonne. And on the top was a note from the editor, telling him why the paper couldn't publish his work. Too many local politicians and businesses would be offended. The mob's role was, and still is, a sensitive topic in Bayonne, and finding that note dealt a sobering slap to my youthful ideals. But some years later *Life* magazine, which could ignore the sensibilities of local merchants and office holders, did a lengthy expose of a mobster named "Joe Bayonne" Zicarelli and his ties to the local congressman, Neil Gallagher. One bizarre and juicy detail: A loan shark named Barney O'Brien dropped dead in Gallagher's house (some say he was in bed with Gallagher's wife), and the congressman summoned a local mob enforcer named Harold "Kayo" Konigsberg to dispose of the body. Gallagher wound up spending more time in federal prison than he did in the U.S. Capitol, and the story reveals one of the seamier sides of Bayonne life. The mob was everywhere and not only Italians were involved. Kayo Konigsberg was Jewish, known by the Yiddish nickname "Heshy" to his family, and his nephew was in my Cub Scout den. That nephew moved to Omaha after finishing medical school to get away from his infamous uncle. The fathers of several friends operated on the edge of the law, and at least one went to prison. During the thirties Abe spent several days in jail for his gambling activities and narrowly avoided other scrapes with the cops. Mom wrote to Dad in Reno that Abe was running a game of some sort in the basement of a synagogue and barely beat a raiding party out the back door.

To a youngster in Bayonne, the most obvious sign of the mob was the numbers racket. Before legal lotteries, organized crime ran a gambling operation that worked this way: You picked a three-digit number and placed your bet with your friendly neighborhood runner, who then recorded the transaction with the bookie. The winner was determined each day by the last three digits of the total amount wagered at a designated racetrack. The odds of winning were 999 to 1, but since the payoff for a successful bet was only 500 to 1, it's easy to see how the mob was minting money. When we moved to Thirty-third Street, I noticed that our local candy store had a large bank of pay phones in the back. At the time, I had no idea what they were for, but eventually I found out. The runners used the phones to call in bets from their territory. A runner was a bit like the Good Humor Man, without the white suit and truck. And he sold dreams, not Dreamsicles. They came true for a lot of people on September 15, 1958, but at a tragic cost.

Investigators reconstructed the events of that morning: At 9:55, the railroad drawbridge crossing Newark Bay between Elizabeth and Bayonne opened for a barge. At 9:57, a commuter train failed to stop and plunged into the water. No one is quite sure why that happened, but the engineer probably had a heart attack. I was just starting my junior year in high school, but since it was Rosh Hashanah, I was in synagogue, not math class. Truth be told, I was more interested in observing the charms of a girl named Bonnie in the next row than I was in observing the holiday rituals. Someone rushed into temple with the news, which the rabbi announced from the pulpit. All doctors were urged to report to the hospital, and half the congregation got up and left. Forty-eight people died that morning, and if not for the Jewish holiday, the toll could have been much higher. One car dangled precariously from the edge of the bridge, and when investigators determined no one inside was left alive, they decided that the safest course was to drop the "death car" into the bay. The next day it was hauled out by a crane, and a police officer took a photo with the car's number, 932, showing clearly on the side. A corpse was visible in the window just above the number. Papers throughout the New York area ran the photo, and hunch gamblers couldn't put enough money on 932. And the next day the pari-mutuel handle at Belmont Raceway was exactly $1,463,932. Area bookies went bankrupt after paying out more than $50 million in winnings.

Italians and Jews were only part of the rich ethnic mix that gave life in Bayonne a special sturdiness and stability. The Catholic parishes served national groups, not neighborhoods, and often used the languages of the Old Country. St. Henry's was Irish, Assumption was Italian, Mount Carmel was Polish, St. Mary's was Ukrainian, and everyone knew where they belonged. The CYO (Catholic Youth Organization) basketball league was a virtual holy war. You weren't just playing for your parish, you were playing for your tribe. A student came to see me recently and mentioned that her family had moved to Bayonne because of the Polish church there. Of course, I said, Mount Carmel. She was stunned I knew this rather esoteric fact, but forty-five years after I left Bayonne, many of these identities remain unchanged.

New ones have been added, however. Churches in Bayonne now serve Arabic, Filipino, Korean, and Latino congregations. A victim of 9/11 who lived in Bayonne was an immigrant from Thailand. Another resident who barely escaped that day was from Egypt. But no Jews from Bayonne were killed. The Jewish community is almost gone. The synagogues in town, once fierce rivals, have now formed a single Hebrew school, but few students are left. Many of the families who use the Jewish Community Center are not Jewish. The mayor, Joe Doria, sends his daughter to summer camp there. And the center's annual fund-raising golf tournament is held at a country club in the ritzy suburb of Basking Ridge.

My brother Glenn tells the story of his first day in gym class at Bayonne High. The teacher asked each kid to say his name, and he dutifully wrote them down: De Martino, Kretkowski, Rosenthal, Demyanovich. When one student answered, "Wilson," the teacher replied: "How do you spell that?" Even if the story is not true, it could be. I randomly picked up the Bayonne High yearbook for 1958 and saw these names on one page: Lombardi, Longobardi, Luikaitis, Lukaszewicz, Machyowsky, Mallamaci, and two Marinellis. There was also a Levin, a Levine, a Levy, and a Lipman. I was immune to the fact that my own homogenized name was ethnically misleading. I was proud of my heritage and I sometimes regretted that my background was not more identifiable. My father might have been worried about anti-Semitism when he changed our name to Roberts, but I can recall only one example of outright, anti-Jewish hostility. Players in the parks department basketball league formed their own teams, and they often broke down along ethnic lines. I was about sixteen, and the Jewish team was vying for the championship against the Polish team. The leading scorer for the Poles, a very large fellow named Stash, was declared ineligible for being too old. So after the Jewish team won our semifinal game and was headed to the finals, the Polish guys jumped us outside the gym. I was one of the top scorers on the Jewish team, and they went right for me. "If Stash can't fuckin' play," they yelled, "you ain't fuckin' playing either. We're going to break your fuckin' arm." I managed to escape with my arm intact, but not my confidence, and in the final game the Polish guys whipped us anyway, even without Stash.

There was a curious segregation about sports in Bayonne. A lot of Jews played basketball and baseball, and dominated the tennis team, but few played football. Maybe their mothers didn't let them. One year we played Weequahic High, a school in Newark that drew from a heavily Jewish neighborhood, the same one made famous by the writer Philip Roth. The name Weequahic is Native American, of course, but I grew up thinking it was Jewish, sort of like Weinberg. After all, the Weequahic Diner was famous for its cheesecake, practically a Jewish national dish. So the football game was a revelation. Who knew there were linebackers named Goldman! Tight ends named Lipschutz! It was like going to Israel for the first time and seeing Jewish soldiers and garbage collectors.

I've always cracked that Bayonne was 80 percent Catholic and 19 percent Jewish, and while those statistics are not quite accurate, I went away to college thinking that Protestants were a tiny minority group. When I arrived at Harvard, I was in for quite a shock. One of the first guys I met was named C. Boyden Gray (he later became White House counsel to the first President Bush), a wealthy and well-bred fellow from all the right schools, with all the right connections. He was a member of the Reynolds tobacco family, his father was President Eisenhower's national security adviser, and he used his first initial. Nobody in Bayonne used their first initial. T. Vinnie Nuccio? M. Paulie Pawlowski? They don't have quite the same ring. I thought Boyden was a Martian. But then, he probably thought I was a Martian, too. My Americanized name did cause some confusion, however. There was another Steve Roberts in my class and he was a blue-blood WASP, a prime target of the elite clubs that still commanded some minor social status in Cambridge and assiduously courted the Boyden Grays of our class. A few of the clubs got the two Steve Robertses mixed up and sent me invitations to various gatherings. I joked with my friends that I should show up and announce loudly, "Guess who's coming to dinner?"

If the parishes in Bayonne were based on ethnicity, so were the politics. The city was governed by five commissioners, and every ticket had to include at least one candidate from each of the three dominant groups: Poles, Irish, and Italians. Then they'd fill out the tickets with various Jews or Ukrainians or Litvaks. One year the principal of our elementary school ran for commissioner, and I remember it clearly because he was so unusual for Bayonne: a strait-laced Protestant who used the title "doctor" and seemed to look down his nose at everyone else. He lost badly. In fact, one Polish politician whose family name was Rudy changed it to Rutkowski so people would know exactly who he was. During my youth three of the leading political figures in town were undertakers, a particularly good way to build a political base in an ethnic community. Their names were O'Brien, Migliaccio, and Fryczynski—one from column A, one from column B, one from column C. In fact, Stanley Fryczynski barely spoke English. It didn't matter, not to his tribe. He was one of theirs. And they deeply revered him for his work in helping Polish immigrants become citizens. During my years as a political reporter, I was often thankful that I came from Bayonne. My boyhood gave me an edge over suburban-bred rivals who didn't know the CYO from the CIO, or a brat from a blintz. This was particularly true during the 1980 campaign, when the so-called Reagan Democrats in the old industrial Rust Belt started moving toward the Republicans and taking the center of political gravity with them. I've always felt at home in cities where smoke-encrusted Catholic churches loom over immigrant neighborhoods, where hand-lettered signs in shop windows advertise homemade kielbasa or kreplach or calzone, where hard hats and work boots are part of the business casual dress code and guys wearing both stop at the local tavern for a beer on their way home from the plant. In Bayonne I learned the importance of loyalty and tradition. I learned what to order in an Irish bar and an Italian restaurant. I learned how to spell Fryczynski, and to understand what a man like that old mortician meant to his community.

But I did get my comeuppance covering a campaign for Congress in Toledo. The Democratic nominee in 1982 was Marcy Kaptur, a young Polish woman whose father was known in the neighborhood for his hand-crafted sausages.

When we met at her campaign headquarters, I asked if there was a café or coffee shop in the neighborhood where we could talk quietly. No, she said, there are only bars around here, but perhaps the priest across the street would let us use the rectory. She dispatched a young aide who returned shortly to say that he couldn't find the father, but he had found another spot for our interview. It was a bit unusual, he said, but very quiet. The Bayonne boy in me burst out. Don't worry, I assured him, I know neighborhoods like this one. No problem. So he ushered us into the front parlor of Urbanski & Sons funeral home. It was indeed very quiet, and very unnerving. Even Bayonne hadn't prepared me for that one.

12

BEYOND BAYONNE

I actually did two things on my first day at Bayonne High. In addition to joining the newspaper, I met a girl. Her name was Renee Cherow, and we were assigned to the same homeroom, 322. She was Jewish, but her family ran a hardware store in the Forties, outside the No. 3 School district, so we didn't know each other. It would be stretching things to call her my high school girlfriend. She was blond and cute and smart, and had many admirers, while my only female fans were related to me. But slowly and fitfully, a friendship developed that became important to both of us as we started to think and dream about life beyond Bayonne.

On Wednesdays many of the high school kids would hang out at the Jewish Community Center, flirting and listening to the jukebox, and on Saturdays there would be occasional dances in the auditorium, capped off with pizza at a joint called Dido's, a few blocks away. I wasn't a regular at these gatherings, since I couldn't dance or even talk very well with members of the opposite sex, but one Saturday I did show up and Renee was there. I summoned all of my courage and asked her to go out for pizza, and much to my amazement, she agreed. Now, Dido's did not exactly offer a fine dining experience. The Cokes were watered down and served immediately, so you had to drink at least two while waiting for your pizza. And when the pie finally came, it was scalding hot, causing second-degree burns on the roof of your mouth if you were not careful. But I wasn't there for the drinks or the food. We sat in a booth, on vinyl-covered seats with a formica table between us. Renee was wearing a gray sweater with a black line across the chest that perfectly accented her blossoming figure. When she leaned over to take a bite of pizza, I think I stopped breathing. I wasn't quite sure what was happening, but I knew it was important. She recalls the sweater but not the evening. Oh, well. I guess it was better for me than it was for her.

I never learned to dance very well, but I did learn to talk, at least with Renee. She, too, was interested in becoming a writer. The *Bayonne Times* asked me to find a girl to help write the high school news column in the Friday paper, and I picked her. We shared an interest in literature and quoted favorite books to each other. One of mine was *Winesburg, Ohio*, by Sherwood Anderson, a series of fictional sketches about small-town life. Winesburg was more rural than Bayonne, and I'm sure the air smelled better there, but the characters and their yearnings were familiar to both of us. Renee and I also wrote poetry, and we both had poems accepted by a volume called *Young America Sings*. Our efforts appeared on facing pages and both were, in a word, dreadful. But our poems, amateurish as they were, reflected the storms of sentiment gusting through us at the time. Hers, called "Sonnet of Remembrance," begins this way:

When music plays and sights and sounds recall
The bygone days of youth, ambition, love;
A melancholia lingers, where with all
My heart I cry, and try to rise above.

Mine, called "First Love," was not so subtle:

O tender one, just once our eyes have met,
And yet my burning love is strong and sure;
The beauty of your face I can't forget;
Your haunting voice sings out so sweet and pure.

I don't know if my poem was inspired by that magic moment over pizza at Dido's. But I do know our friendship was. After that, we grew more comfortable with each other. We could share our secret hopes. We could say to each other, without being laughed at: There's a big world out there, a world where we can live and love and write. When I asked her years later to describe our relationship, she summed it up by saying, "Words." I think that's right. We reflected each other, reassured each other, practiced on each other. On nice evenings we'd sit on the front stoop of her house on Avenue B, or on a bench in a nearby park, looking out at the bay. When it got wet or cold, we'd sit on the floor of her front hallway. When we progressed to occasional and awkward necking sessions, we'd sneak upstairs into her front parlor. Like our house on Thirty-first Street, hers was a two-family dwelling, with the Cherows occupying the second floor. And Renee, too, had a grandfather living with her, who would appear suddenly in the hallway, looking for the bathroom and causing us to spring apart on the couch. And that's eventually what happened. We sprang apart, attending different colleges, pursuing different careers, marrying different people. But when I hear the music of "bygone days of youth," I think of her, and what she meant to me back then. She was the first girl to take me seriously. Maybe, even, to think I was special. I know I thought she was special. And it all started over tomato sauce and mozzarella cheese.

Clan elders in Bayonne were happy to have their youngsters mix with other religions on the playing field, but the dance floor or the pizza parlor—to say nothing of the living room couch—was another story. And Catholics and Jews both created institutions designed to deter romance across tribal lines. Many of the brightest Catholic kids were siphoned off to Catholic schools in the region, and while that worked for the church fathers, there were side effects for the rest of us. This meant that in the only public high school, Jews dominated the advanced classes and the honor societies, leaving us with an unhealthy and inaccurate view of our own superiority. A few Orthodox Jews, like my basketball buddy Barry Eisenberg, also went to religious schools, but the real effort at social segregation came through the Jewish Community Center. I played basketball there, went to Boy Scout meetings there, served as a summer camp counselor there. And the system worked. There were few marriages across religious lines in Bayonne, and not all that many across ethnic lines, either. Those Donnelly girls on Thirty-first Street, marrying an Amato and a Pawlowski and a Seestead, were about as bold as anyone ever got.

Only a few years later, however, the lines started blurring. A student of mine, call her T., told me the story of her parents: Her Jewish father and Italian mother met at Bayonne High, and when they started dating, her mother's parents were so incensed they threw her out of the house. T.'s mom moved in with her future in-laws, married the guy, and converted to Judaism, but she retained a strong loyalty to her Italian origins. So T. and her mom have a ritual: Whenever they go shopping in New York, they stop at St. Patrick's Cathedral and light a candle for their Catholic ancestors. And sitting there in the cathedral, T. says Hebrew prayers, the only ones she knows. I think of it as a very Bayonne story.

But clearly, even in high school, I was intrigued by Catholic girls. I'm sure the "forbidden fruit" allure was part of it. More to the point, I had known many of the Jewish girls forever, and they had known me. I had felt uncomfortable around the Barrie Posnaks and Lois Turtletaubs since third grade, and that basic relationship wasn't going to change. With the Catholic girls I could start fresh. My braces were coming off, my pimples were clearing up, I grew five inches in one year, and they hadn't known me when the other boys would steal my hat and make me cry. A girl named Joyce Balchunas invited me to her sweet sixteen party at the Polish-American Home, and I would occasionally walk her home. One day we started kissing on her front steps, and I suggested that we continue around the side, in the alley. I made up some excuse, but the real reason, I'm ashamed to admit, was fear. I was afraid someone I knew, someone Jewish, would see us. As for Gigi Barresi, I never did get around to kissing her—to my everlasting regret, I might add. Our relationship was confined to chatting in the hallways of the high school or on the sidewalk out front. I don't know how the flirtation started, but we were both intrigued. And intimidated. Her locker was near the cafeteria and I found excuses to run into her. Her family owned a dry cleaners down on Avenue C and I once drove by, hoping I'd catch a glimpse of her, but I never had the guts to stop the car, let alone ask her out. Not even for pizza. The taboos were just too strong. As graduation approached we did exchange pictures, and it's the only one I've saved of a high school classmate, tucked away in a cardboard cigar box. There she is, short dark hair, shining dark eyes, a smile so bright you could light a votive candle from it. On the back she wrote: "Steve, Promise never to use this picture as a dartboard and I promise never to use yours. Do you realize we were never introduced? Gigi. P.S. I almost forgot.

Lots of Luck in anything you do." Well, I kept my part of the bargain, Gigi. Forty-five years later and there's not a single hole in your picture.

The taboos were so powerful because they were so deeply rooted in the culture of Bayonne. My parents had few if any non-Jewish friends, and I didn't, either. I played ball with guys like Moose Poplowsky and Eddie Broderick, but we never talked about anything important or saw each other off the court or away from the diamond. Bayonne is a small place, but when I drove to Gigi's store, or walked to Joyce's house, it was like crossing a frontier into another country. So when I went off to college, and met a Catholic girl, and started bringing her home, my parents were pretty upset, to say the least. Like most people, they feared what they didn't know. But Dad had a deeper reason. His old anxieties returned. No Jewish man, he told me, married to a Catholic woman, could ever be the "dominant male" in the family. He had long been convinced that one twin would never find love because his manhood was physically impaired. And now the other twin had found love with the wrong woman, a woman, he worried, who would leave his manhood psychologically impaired.

That battle, however, was in the future. For now, I was still in high school, and all those small steps—talking with Renee, flirting with Gigi, necking with Joyce—were leading me in one direction, beyond Bayonne. I took plenty of tumbles along the way. Occasionally, the sports editor of the *Bayonne Times*, a man universally known as Rosie, would give me tickets to Yankee games, and I'd take the bus into New York, getting off at the Port Authority, which was just a few blocks from Times Square. On one trip I decided to buy a tie at a shop on the square. I'm not sure why, I didn't own one and didn't need one, but my after-school job provided me with some pocket money, and I must have spent an hour pondering every tie in the store. I'm sure the clerk thought I was either a shoplifter or a deeply disturbed silk fetishist. Finally, drenched in sweat, I made my selection—a black knit tie, probably the plainest one in the entire place, well suited for funerals and not much else. Until I got to college. In those years the dining halls at Harvard required men to wear ties at all times, and that black knit number saw me through hundreds of meals.

At about this time, Dad found, hidden away in a drawer, a gift certificate from a fancy New York men's store, A. Sulka. It was years old, he couldn't even remember where it came from, but he suggested I cash it in on my next trip to the city. The face value was $25, as I recall, almost two weeks' wages for me at the *Bayonne Times*, and the folks at Sulka were rather appalled when I showed up in their store, the yellowing document clutched in my clammy hand. Reluctantly, they agreed to accept it. What would you like? the clerk asked me, not bothering to disguise his disdain. Since the only tie I had ever bought cost $5, I thought I was safe in my answer: two ties. I was quickly corrected. You mean one tie, sneered the clerk. By this time my confidence, not robust to begin with, was totally flattened. I quickly picked out a tie, any tie, threw in a pair of socks and got out of there. I don't think I ever wore the tie, and I've never been back to Sulka since.

I played on the Jewish center's basketball team, and on Sunday afternoons we'd play teams from other towns around the state. Many of the centers, like ours in Bayonne, were new buildings, proudly financed by professionals and business people enjoying the fruits of postwar prosperity. Some were older, and the center in Trenton provided the ultimate lesson in home-court advantage. The basketball court was on the second floor, a converted auditorium, and there were windows right on the edge of the court. Sure, there was wire mesh over the glass, but you were always careful not to dive for a ball on that side. Even worse was the ceiling. On one end of the court it slanted right down to the top of the backboard. With long years of practice, the hometown players knew exactly how to arch their shots to avoid this obstacle. We didn't. The ceiling acted like a goaltender, swatting away our best efforts. And besides, we were petrified of plunging through the windows to the street below. I don't think Trenton ever lost at home. I dreamed recently about returning to Trenton and playing those games again. But glory eluded me, even in my fantasies. My aging knees hampered my jump shot and we still lost.

Playing ball at Jewish community centers was easy—I wore a uniform, I had a role and a number, I knew what I was doing. Picking up girls was much harder. Besides, you couldn't drive until you were seventeen in New Jersey, and since we had started school young, that meant no wheels till the middle of senior year. But when we were fifteen and sixteen, Marc and I started hanging out with a group of older guys who thought of themselves as intellectuals. And in fact they were. The leader of the pack was George Lakoff, now a distinguished linguist at the University of California, who lived on the next block. We formed a club, called the Hedonists, a name suggested by Dad, and reveled in the rather snobbish conviction that no one else in Bayonne knew what the word meant. We often gathered at our house, playing cards on the kitchen table for matchsticks, unconsciously copying the games Grandma Miriam had once presided over. We formed a softball team that we called Izzy's Fish Market, a name inspired by a high school math teacher, Isadore Chertoff, who would frequently remind his wayward students that he was running an academic class, not a fish market. When Neil Miller, another pal, could borrow his father's milk truck, we'd sometimes drive to the Jewish centers in Englewood or Ridgewood or Maplewood—or one of the other Woods—in search of girls who didn't know we were the biggest nerds in Bayonne High. In fact, I soon realized that any town with a geographical landmark in its name—River Edge, Short Hills, Westfield—was probably a lot ritzier than Bayonne or Newark or Jersey City. And prospecting for female companionship in those leafy locales while driving a milk truck really cut down your chances. After all, there was only one seat beside the driver, and everyone else had to sit on milk crates in the back. In fact, I don't think we ever lured a girl into riding in the back of that truck, but we kept the faith. We'd end up most nights at a local diner, or at our favorite drugstore, drinking milk shakes at the counter and telling each other that next time—next time—we'd be more successful.

Steve Davidowitz put our milk truck to shame. He was a pitcher on the high school baseball team, a hard-throwing lefty who was just wild enough to keep batters antsy. And one day, he disappeared. No one could find him. When he finally surfaced, we learned the story: He had taken his bar mitzvah money from his bank account and flown to Cuba, apparently lured by the bright lights and fast company of Havana's high life. This was absolutely the most daring, dramatic story any of us had ever heard. When he returned a few days later, Steve became an instant folk hero, and of course, forever after, was known by the nickname "Cuba." Apparently he was like me, drawn to his lifework at an early age. He became an expert on horse racing and author of a classic in the field, *Betting Thoroughbreds*.

I never stole my Bar Mitzvah money or flew to Cuba, but my travels beyond the borders of Bayonne opened my eyes in other ways. Marc and I got involved in an organization called the Jersey Federation of Temple Youth, "Jifty" for short. It brought young people together from Reform Jewish congregations for weekend "conclaves" that were equal parts religious seminar and mating ritual. Like our pals in the Hedonists, the people we met through Jifty actually read books and argued about issues. One of our buddies, Peter Knobel, became a rabbi, and another, Alan Kors, a professor of history. Our late-night bull sessions at those weekend meetings gave us all our first taste of college life. Without the beer. And the dances and parties gave Marc and me a chance to remake ourselves or, more precisely, to be ourselves. The process of "starting fresh," which began with Gigi and Joyce back in Bayonne, blossomed with the Ina Friedmans and Myrna Goldblatts I met through Jifty. When I talked, girls actually listened and laughed. They had no idea what I used to look like or act like. It was a giddy feeling. Probably close to what Willie Rogow was feeling when he told funny stories at Dotty Schanbam's seventeenth birthday party and she let him walk her home.

So I had nothing but good memories years later, when a

Marc and me (with three young ladies) outside Temple B'nai Jeshurun in Newark during a Jewish youth group meeting in 1958.

woman called who had been at Bayonne High a few years ahead of me. She now ran the speaker series at Temple B'nai Jeshurun in Short Hills, New Jersey, and would I come? I was thrilled. The first conclave I ever attended was at B'nai Jeshurun, then located in Newark. The congregation had left the city years before, sold its building to a black church, and relocated in the suburbs. But the rabbi, and many of the families, were still the same—including the parents of my old friend Peter Knobel—so I immediately accepted her invitation. The evening arrived. My parents were so excited they attended the speech and brought several friends. I got up and recalled my past. Speaking at Jeshurun was like coming home, I said. I used to attend conclaves at the old building in Newark, and I remembered it with great fondness. In fact, I told my audience, "I remember when most of you still lived in Newark." Dead silence. No one wanted to be reminded of their origins. My parents were mortified in front of their friends. So much for sentiment.

Midway through senior year, I finally got my driver's license, and Dad got us a rattletrap of a car to tool around in. Perhaps because of his own predilection for running out of gas, he always assured us: If you have a problem, call me and I'll come get you. And at least once, he had to make good on that promise, rescuing us in darkest Irvington after some vehicular mishap. Many years later, when my son incurred two flat tires on his way home from college, he reluctantly called in the middle of the night, asking for help. When I finally reached him, and we were heading home, he admitted how hard it had been to make that call. But he didn't have a choice. Believe me, I replied, relieving him enormously, I understand, I've had to make those calls myself. In fact, during my first trip outside Bayonne, to some diner in Jersey City, I was dismayed to realize that no one had taught me how to park in a parking lot, only on a street. We didn't have many parking lots in Bayonne, and I didn't come close to making it between the white lines. Still, having a license and a car, no matter how decrepit, was a passport to a wider world, and one moment stands out. If you wanted to leave Bayonne heading west—toward the Green Paradise of Woods and Ridges, Inas and Myrnas—you had to drive across a bridge spanning the upper reaches of Newark Bay. On the first trip I drove by myself, crossing that bridge to some weekend meeting, I felt like I was flying off the end of the earth. And I was.

WHERE'S HARVARD?

Sometimes, a five-minute conversation can change your entire life. I had one of those during my senior year in high school. At the Jewish Community Center one night, I was approached by a fellow named Barney Frank. He was a few years older, but I was friends with his younger sister, Doris, and Rosie, the sports editor who gave me passes to Yankee Stadium, was his uncle. Barney's older sister, Ann, had gone to Radcliffe. He had followed her to Cambridge and was now at Harvard, and he suggested that Marc and I should apply. I was not quite sure where Harvard was, but he made it sound like a neat place; so on a lark, we both filed applications. Marc also visited Princeton, but had an off-putting experience. The student guide told him flatly that Princeton still had a strict Jewish quota, and he shouldn't get his hopes up. So he lost interest.

I had never flown in a plane, and this was not the time to start. I had barely been out of New Jersey before, and going all the way to Massachusetts for a college interview was more than enough stress. So Marc and I took a train from Newark to Boston for our first visit to Harvard. It was a bit like Abe leaving Russia for Palestine—we were sixteen, the same age he had been when he first left home—but we didn't have to steal money from our father to finance the trip. When we got to Boston, we followed Barney's instructions and made our way to the MTA, the local underground transit system. Next thing I knew, this odd-looking vehicle rolled into the station. I was used to New York City subways, and I had never seen

I'm making a speech at Albertus Magnus College in 1962, not long after meeting Cokie.

anything like it. I thought a bus had made a wrong turn and wound up in the subway tunnel. We got on anyway, and made our way to the Harvard Square stop.

Emerging on an island in the middle of that busy intersection, I walked over to a man running a newsstand and asked timidly, "Where's Harvard?" He looked at me kindly and replied: "It's all around you, son." We were feeling pretty lost when we spied, walking across the street, none other than Barney Frank. What a relief! (His companion that night, Goldie Rosenhan, a student at Brandeis, was the woman who asked me, years later, to speak at the synagogue in Short Hills.) We stayed in Barney's room in Kirkland House, but were disappointed at the lack of a real, intellectual college bull session. When I mentioned this, Barney pointed to his two roommates and laughed: "We've been roommates for three years. Whatever there is to be said about politics has been said." (Not quite, it turned out. Barney went on to a highly successful career in politics and today represents Massachusetts in Congress.) We were interviewed the next day by an admissions officer who had spent the war at the Bayonne Navy base where Abe had cadged all those extra plywood planks. The fact that he knew Bayonne relaxed me a lot. We had something to talk about. And that spring Marc and I both got letters accepting us into the class of 1964.

I was never quite sure why we both got in, but during our freshman year, we learned that a Harvard teaching hospital was doing a study of twins, so perhaps they wanted us as subjects. In fact, the hospital did invite us to participate, and I was intrigued by the idea, except for one thing: They wanted us to save all of our urine in large plastic bottles. I had enough trouble getting a date, I didn't need to worry about a fluid-filled container that might leak at any moment. We took the bottles home, but never joined the study, and throughout my college years, those damn things gathered dust under my bed, a mute reminder of my social insecurity.

The fact that we could even apply to Harvard and contemplate going there reflected Dad's improving financial fortunes. His departure from publishing had been followed by long years of scarcity and struggle. Just a year before we applied to college, he sat us down and told us frankly that we had to get scholarships or attend a state school. He just couldn't afford anything else. Then he finally caught a break. Working for his father had been a terrible experience, but he'd learned a lot about the mobile home business, and when he heard about an opening for a trailer park manager in northern New Jersey, a reasonable drive from Bayonne, he applied. Then he went out and bought a new car he couldn't really afford, a blue Dodge with tailfins, in order to make a good impression. It worked. He got the job, and that car became his good luck charm.

If Dad's irregular work habits had caused him problems as a young man, he plunged into this new opportunity with enormous energy, and sending us to college was one of his major motivations. He drove himself hard and turned the park around. Then he started selling mobile homes as well, and discovered he was a gifted salesman. By the time we got those acceptance letters, Dad and Mom could finance our education by themselves. That was always a source of great pride to them, and rightly so, but it came at a cost. I never think of Dad as a negligent father, but he wasn't home a lot in those years, and he once said to me with some pain: "One day, I looked up and you were gone." He didn't mean to impart a lesson, but I took it as one. I promised myself I would never feel that way about my own children.

Dad continued to manage the trailer park through our college years and then, in 1965, grasped an even better opportunity. The builders of a new park in Lakewood had run out of money and were selling at a bargain rate. Dad borrowed to the hilt, took a chance, bought the place, and renamed it Roberts Park. Lakewood was the same town where Abe had sold all those boards to local chicken farmers, and where I had once been held captive by an irate rooster, much to the amusement of my family. My parents left Bayonne for Lakewood, where the children of those chicken farmers became some of their closest friends. The gamble paid off. The park prospered. Dad was not publishing books, or writing them. But the man who had returned from Reno in 1939, unable to afford a thirty-five-cent telegram, could now give his children a great gift, a measure of financial security. And he did it with his own hands—leveling concrete, digging trenches, plowing snow, hauling trash. When he was his own boss, no one worked harder. It's no accident that Dad was happiest—and most successful—running the two businesses that bore his own name and stamp: Will Roberts Publishing and Roberts Park. Like his father, he was a solo act, not a sideman, and when he wanted to, he could really play. About that blue Dodge: When it finally died, Dad actually buried it on a back lot in his park. We even have pictures of the interment. He didn't want such an important talisman of good fortune consigned to a junk heap.

When we got our letters from Harvard, Marc announced immediately that he was going. So I announced that I was not. Since birth, we had always been thrown together—in the same bedroom, the same classes, the same organizations. Sure, my athletics and journalism set me apart in some ways, but to most people we were still the "twinnies" and I was tired of it. The competition between us certainly had a beneficial effect, forcing me to work hard and get good grades. Without his example, I would have shot a lot more baskets and cracked a lot fewer books. But I didn't want to be "Marc Roberts's brother" anymore, and one incident summed up my feelings. Marc was chosen to represent Bayonne High at Boys State, a prestigious gathering of politically minded high school students from across New Jersey. I was not selected, which was bad enough. Then he got elected governor, an honor that entitled him to attend Boys Nation, a meeting in Washington of all the governors from other states. That sealed it for me. No Harvard.

I realize now that we both had issues with each other, that we both felt overshadowed by our twin brother. I envied his titles and achievements. He envied my friendships with girls. For all my awkwardness, I had it a lot easier than he did, and he lusted after Renee from afar. So perhaps we both would have preferred separate schools. My parents played it cool, letting me simmer for a few days. They had no family ties to Harvard, of course. Bialystok and NYU and the Workmen's Circle yes, Harvard no. But they did have a friend who had gone there, and she told them strongly: Don't let Steve miss this chance, he'll always regret it. Finally, Dad came to me and said, Look, the place is big enough for both of you. I was still reluctant, but Harvard was the best college that had accepted me, so I grudgingly gave in. Very grudgingly. Once the decision was made, all I had left to do in high school was finish a senior research project for a wonderful teacher of mine, Mollie Bayroff. It was the spring of 1960, I decided to write about the fight for the Democratic presidential nomination, and I spent hours in the public library, reading magazines like the *Reporter* and getting my first real taste of political writing. Like all good liberals, I favored Hubert Humphrey and predicted he would defeat Jack Kennedy, whom I described as a "bully boy from Boston." Obviously, my analytical skills needed work. But that would have to wait for college. Graduation came. We walked down the aisle as the band played our school fight song one last time: "Bayonne goes marching down the line, everyone is feeling mighty fine . . ." But I wasn't feeling fine, I was feeling petrified. I wasn't ready for Harvard. I would never fit in. It was a crazy idea. And I didn't know the half of it. As I picked up my diploma in June 1960, the world was taking a sharp swerve into a new decade, but I couldn't realize what was happening. None of us could. We were children of The Fifties set adrift in The Sixties. And we were totally unprepared for the tidal wave that was about to swamp our little boat.

Ten years later I returned to Bayonne for my high school reunion. I was then the *New York Times* bureau chief in Los Angeles, specializing in antiwar protests and counterculture experiments, and I wanted to see how my high school classmates were adapting to all the changes crashing around us. The article I wrote for the *Times* Sunday magazine, "Old-Fashioned at 27," expressed how a lot of us were feeling, and how a lot of readers were feeling. But I also said some things about Bayonne that stirred deep resentment in my hometown. The result was a torrent of letters, pro and con, the biggest response I ever received during my *Times* career.

I started by describing the city: "When some people think of home they remember tree-shaded streets or a particular park or maybe a river. I think of the soot of Bayonne and its smell, an aromatic combination of barbequed garbage and smoldering inner tubes, with a dash of sulphur for tanginess." As I watched the crowd gather at the Hi-Hat Club—the same club Jackie Gleason and Rodney Dangerfield had once played—I reflected on our high school years: "It was a marvelously uncomplicated four years. Grass was what the sign said not to walk on in the county park. Sex was something that happened to somebody else. Some of the older guys carried condoms around 'just in case,' but they usually crumbled from age and disuse." New York was just across the Hudson River, but the barrier between us and "the city" could have been the Berlin Wall: "We heard vaguely of 'beatniks' and 'foreign movies,' and maybe we braved Greenwich Village for a cup of cappuccino, but for all we knew, Jack Kerouac could have been a third baseman for the Dodgers." The questionnaires they filled out indicated that many classmates had been married six or eight years, some as long as ten (I had been married four years at that point). "Most of them have decent, average jobs," I wrote, "salesman, dentist, insurance agent, accountant, hair dresser, office clerk, mortician, computer programmer, airplane pilot, fireman, toll collector on the New Jersey Turnpike." Under "places traveled," a few said Europe, many said Florida, only two mentioned Vietnam, a sign that most of us were a bit too old for the war. One mother of three replied to the travel question: "Are you kidding?"

Then I tried to sum up the difference between my life in Los Angeles, and the one I had left behind in Bayonne: "Because I went away to Harvard and have the kind of job I do, many of my friends are strivers, achievers. They want to have an influence, preferably on a national scale. What's more, they feel they deserve regular doses of passion, of intense experience, whether through drugs or art, sex or skiing. They want—or at least talk about—life with a capital L. Not my high school classmates. They are neither ecstatic nor depressed, and they do not expect to be. They are neither movers nor shakers, and that is all right. They are content. They get along."

Mike Lipman, a physics teacher at Bayonne High,

summed up the tone of many conversations that night: "I think we just missed out. The last five years have really been the turning point." Mike was right, we had just missed out, and the result was a continuing conflict, between the values we had been taught and the culture we couldn't escape. "I have to get back to work," said Joyce Frank Fishberg, who had quit her secretarial job in New York to have a baby. "I like women's lib, but I've been brainwashed by the old school—I'd still feel guilty if I left my child." Pat Blihar Connors was also old school. Her husband, Brian, was a cop, and her mother lived across the street, an "excellent babysitter" for her three little kids. She worried about Brian: "I don't like my husband being called a pig"; but even more about her kids: "I hope I'll be able to meet their needs as they get older. Each day is another problem, and I only hope I can handle it in the right way, especially with values changing so rapidly. I'm kind of old-fashioned—not puritanical but old-fashioned—even compared to a 21-year-old. Sometimes being a mother is a very awesome thing." After listening to Mike and Joyce and Pat and the others, I ended this way: "We had been shaped in the dying years of a world that no longer exists. We had been taught that you play by the rules, that you respect authority, that reason is the source of all wisdom, that caution is a virtue, that sex is dirty and that a woman's place is in the home. We could just as easily 'let it all hang out' as fly to the moon. At 27, we are old-fashioned. And that is not a very comfortable thing to be these days."

When I filed the piece to the magazine, my editor called

back quickly. We love it, he said, but I guess you don't want to go back to Bayonne anytime soon. I was stunned. What are you talking about? I love Bayonne! Trust me, he replied, they'll hate it. And they did. The *Bayonne Times*, which had been so proud of me, wrote an editorial, called "Steve Roberts' World": "Bayonne does not come off well in the Roberts article. Only Steve himself comes off worse. We can only hope that he is more aware of the world about him in California where he reports from than about Bayonne, where he lived for 17 years. Otherwise New York Times readers had better discount his stories." Even Grandpa Abe, who still had a business in Bayonne, wrote to the paper and put his message in the form of a letter to me: "Fifty years ago, I came to Bayonne on a motorcycle with my carpenter's tool box in the sidecar. In no time at all, I started doing things." He listed all his accomplishments on First Street and described adding "five acres of land out of water in Newark Bay" to build his trailer park. "It has been fifty years of excitement and satisfaction," Abe went on. "When I came to Bayonne, the proverbial European soap was still wet behind my ears. Yes, Steve . . . if I had it to do all over again—I would pick Bayonne again. To me it is the greatest city in the U.S.A." He signed it: "Grandpa Abe Rogow, 31 Lincoln Parkway."

Outsiders said the piece captured their own feelings of confusion and ambivalence, but people from Bayonne generally agreed I was a sneering snob looking down on my hometown. And looking back from my current perspective, I see their point. I tell my writing students that if readers misinterpret your meaning, it's your fault, not theirs, and so the blame for any miscommunication is mine. In talking about the life I was living in L.A., in describing the "strivers" and "achievers" I knew, I was trying to compare them unfavorably to my high school classmates, who struck me as much more levelheaded and well grounded. Apparently that didn't come across, at least not in Bayonne, and when I attended another reunion a few years later, people were still so angry that they practically frisked me at the door, to make sure I wasn't carrying a concealed notebook. But my admiration for my classmates, and my hometown, has only grown over the years. Yes, we felt overwhelmed by the sixties. We wondered whether the "old-fashioned" values we had learned in the neighborhoods and around the kitchen tables of Bayonne would stand up. Today, it's clear to me that they have, with the glaring exception of our attitudes toward women and sex. Those values were the rock we built our lives on. We're old-fashioned and proud of it.

That's how I feel now. In the spring and summer of 1960, I was clueless and confused. I knew I wanted to break away, to do this college thing on my own, but I had no idea about what that meant. Take the shirts. Mom went to a factory outlet in Jersey City and bought me—at a nice discount, of course—an armload of new shirts to take to college. Unlike the shirts Dad took to college, they were not identical. But I blew up at her. Don't you understand I want to buy my own shirts for college, I screamed. But if you had set me loose in a shirt store, I would have been paralyzed. It took me an hour to buy one black knit tie. Shirts would have taken days.

Then there was the meeting in New York for incoming freshmen from the local region. A national math contest was held every year for high school seniors, and the highest scorer in each school received a pin. Marc won the pin for Bayonne High and wore it proudly to the meeting. Another fellow was wearing one, too, and they fell into conversation. What score did you get? the other guy asked Marc. He answered with a figure near 100 out of a perfect score of 150. What about you? The answer was 148 or 149. He'd gone to Bronx Science, and had one of the highest scores in the entire country. These were the students we'd be competing against. Yep, this was a mistake.

About this time I was getting to know Lisa, the girl I met on Madison Avenue. Somehow, our disastrous dinner, featuring that chilled and naked Chianti bottle, didn't end our relationship. Perhaps she knew how much I needed help, and took pity on me. So at the end of the summer, she invited me to have a drink with her father, an old Harvard grad then working in public relations. The idea was to give me a feel for the school, to answer some questions, and put me at ease. But the opposite happened. First, I didn't know the name of a drink to order. I'd never had a real drink, let alone ordered one at the Harvard Club. In our house a bottle of schnapps, the Yiddish word for whiskey, could literally last for years. Finally, I requested a Coke, and Lisa's father struggled to disguise his disappointment. Where did you get this guy? The talk turned to Harvard life. Every undergrad is assigned to a house after freshman year, and in that era they each had a distinct personality and reputation. But I had only read the names of these places, I had never heard them pronounced. I casually mentioned that Winthrop was the jock house, but I said it "Win-THROP," emphasizing the second syllable. That, corrected my drinking companion, is pronounced "Winnthrripp," with no accent and a barely discernible second syllable. Yep, this really was a mistake.

The fall came, Mom and Dad drove us to school. I was assigned to Holworthy Hall, named for a wealthy seventeenth-century British merchant, and all the other residences had similar pedigrees—Weld and Wigglesworth, Thayer and Hollis. Not a Rogowsky or a Barresi in sight. That first day, wearing one of the shirts Mom had bought, I saw three guys from the floor above walking down the stairs toward me. They wore identical costumes: Madras Bermuda shorts and loafers without socks. And they were smoking pipes. Now, I didn't even own any shorts or loafers—having been brainwashed all those years about the virtues of shoes with laces—and going out in public with no socks was as scandalous as going out with no underwear. On top of that, I couldn't keep a pipe lit. I knew, because I had tried one of Dad's during the summer. No dice. Yep, this was a really big mistake.

That night, the proctor, as the dorm supervisor was called, gathered all the freshmen under his command for a chat. It didn't go well. Start with his name, Bevis Longstreth. He was a law student, the scion of an old and wealthy Philadelphia family, and there was nobody in Bayonne with a name like that. Plenty of Longobardis, but no Longstreths. I was sure he owned many pairs of Madras shorts and at least a dozen loafers. Then he served beer. Now, the prep school guys thought this was great. They cracked open their cans and took giant gulps. I had never drunk beer before. Ever. I dutifully opened a can and took a sip. Ugh. Awful. Soon the preppies had finished their first cans, tossed them ostentatiously into the trash barrel, and opened a second. My can was still mostly full. I briefly contemplated calling on my old basketball skills and artfully sinking a jump shot into the barrel with my beer. But I was too scared. What if I missed, and the can rolled around on the floor, spewing foam? The chance for humiliation was too great. I stuck with my first can.

As I sat there, nursing that beer, and listening to Bevis explain the rules, I really was that Chanti bottle I had shared with Lisa, wobbly and weak in that bucket of ice. The support system that had kept me upright—my parents and grandparents, The Block and the *Bayonne Times*, the Elks Club and the Jewish Community Center, Renee and Gigi, the Inas and the Myrnas—they had all been stripped away. I didn't own the right clothes or know the right drinks, and I couldn't even pronounce the names on the buildings correctly. But I was a Bayonne boy. I knew who I was and where I was from. And even if I didn't realize it yet, I was ready to become a Harvard man.

THE PERFECT GIRL

Somewhat to my surprise, over the next four years I did become a Harvard man. I never learned to smoke a pipe, or swallow more than a single beer at a time, but I did come to feel at home. After growing up in a house my grandfather had built with his own hands, I moved into a dorm erected a hundred years before my family arrived in America. Yet the two buildings had something critical in common. They were both filled with books and papers, arguments and ideas. They were both places of learning. I left Cambridge more than forty years ago and took on many different roles: newspaper reporter and TV commentator, columnist and professor, husband, father, and grandfather. But those identities all tie together. They are all part of the same tradition, a tradition I was taught in my fathers' houses.

Still, I had to adjust to a new life in a bewildering new

Cokie and me as young marrieds, not long after I was posted to California as Los Angeles bureau chief of the New York Times *in 1969.*

place. Two months after I arrived at Harvard, John F. Kennedy was elected president, and in January, before taking office, he came to campus for a meeting of the Board of Overseers. Our front window on Thirty-first Street looked out on The Block, the scene of stickball games and Good Humor trucks. Our front window in Holworthy Hall looked out on The Yard, and from that perch I could watch Kennedy stroll to his meeting. What I couldn't know, at that moment, was that I was taking part in the great American experiment in democracy. My grandparents were new immigrants, not old money. They hammered in nails for a living instead of hammering out deals. They spoke rough-hewn Yiddish to their *landsmen* at the Labor Lyceum, not high-toned English to their classmates from prep school. I played softball in the schoolyard, not squash at the country club. I had a taste for Dreamsicles, not daiquiris. Yet here I was, a few yards from the new president, studying at the same college he had attended, and no one seemed to care about my grandparents' origins or accents.

Sure, Harvard was an elitist place, and I had plenty of well-born, well-connected classmates. But even the most exclusive American institutions are also open to outsiders like me, whose family influence could get you a good deal at Uncle Emil's watch repair shop on the Lower East Side and not much else. Down the hall lived the son of Arthur Schlesinger Jr., the noted Harvard historian who would soon join Kennedy's White House staff. The pinnacle of my father's political career was electing a high school drafting teacher as mayor of Bayonne. However, in a basic sense, young Schlesinger and I were equals. We both had the same chance to make our mark on campus. And there was only one place for me to do that, the *Harvard Crimson*.

From the moment you walked into its somewhat seedy headquarters at 14 Plympton Street, you were reminded that "The Crime" was not your usual college newspaper. A large wooden chair in the "sanctum," a room on the top floor, was adorned with plaques listing the paper's previous presidents. If you waded through the discarded pizza boxes and pop bottles, you could read that one of them had been Franklin D. Roosevelt, Class of 1904. If that wasn't intimidating enough, you had to survive a competition, or "comp," to get accepted as a staff member. It was a form of hazing, meant to discourage the casual or uncommitted journalist, and the washout rate was high. But I had an advantage. I already knew the aroma of stale ink and hot lead that had seeped into every crack and corner of the *Crimson* building.

I signed up for the first comp of the fall, and one of my earliest assignments was to interview the college dean, McGeorge Bundy, about a new building then going up near Harvard Square. He sat at a big desk, at the back of a big office, on the top floor of the same building where Kennedy would attend the Overseers meeting. As I walked gingerly toward him, Bundy fixed me with a hard stare, and the room felt like it was the size of a football field. When I was finally seated at the very edge of my chair, he asked if I was competing for an executive post at the *Crimson*. No, sir, I answered, I'm just a freshman, I'm competing for the news board. He growled with impatience: "Then how the hell did you get in to see me?" His act had the intended effect. I was paralyzed. But only for a moment. After Sandra Dee, I could handle Mac Bundy. Gidget was a Star, Bundy was only a Dean. I got through the interview, and the comp, and on December 13, 1960, my father's forty-fourth birthday, a small box on the *Crimson*'s editorial page announced that "Steven V. Roberts

'64 of Holworthy Hall and Bayonne, N.J." had been elected to the news board. (Anthony Hiss '63 of Adams House and New York City was elected to the editorial board on the same day. He later wrote a book about his father, Alger Hiss.) My first bylined story a day or two later was hardly memorable, a preview of a basketball game against Tufts University, but soon my name started appearing regularly in the paper, and one day someone greeted Marc by saying, "Oh, you're Steve Roberts's brother." At that point I knew Dad had been right, the place was big enough for both of us.

The *Crimson* became my fraternity, my classroom, my social center—the place where I belonged. I knew something about newspapers, but very little about the world they covered, and one evening I was just hanging out, listening to a couple of older guys talking about the war in Algeria. I literally did not know where Algeria was, but now I had a reason to find out. In fact, the *Crimson* soaked up so much time and energy that my class work faltered. I was never very good in languages, and after I almost failed French during my first midterm exams, I called home in distress, vowing to quit the paper and devote more time to my studies. Dad let me vent for a while and finally said, Don't be ridiculous, you'll get a lot more out of writing for the paper than studying French, don't think about quitting. He was right again. Years later I found his college record and was thrilled to see all the Ds he'd earned in French.

That was a heady year to be in Cambridge. Half the Harvard government faculty joined the Kennedy administration in Washington, but the campus seemed fixated on an issue of less cosmic importance: Should college diplomas remain in Latin or be converted into English? President Kennedy, still a member of the Board of Overseers, would have a vote on the matter. And since our old dean, Mac Bundy, had become the president's national security adviser, we decided one night, in a burst of youthful arrogance, to give Bundy a call and find out Kennedy's thinking on the matter. Then the older guys chickened out. So they deputized the raw rookie from Holworthy Hall and Bayonne, N.J., to try and reach Bundy. Somehow, I got through, and when I told Bundy I was calling from the *Crimson* he sneered, "The *Crimson*? Are you guys still in business?"

Just a week or two before, the Bay of Pigs invasion had ended in failure, and the first criticisms of the new president were just starting to appear. So at the end of our story, we quoted Bundy's line about the *Crimson* still being in business and added a final wisecrack: "The Crimson, in a fit of self-control, refrained from asking Mr. Bundy the same question." The national press picked up the story and had a field day. Here was Kennedy's own college paper, slapping him around. Of course, back in Cambridge, we were loving every minute of it. But we hadn't heard the last from Mac Bundy or Jack Kennedy.

We sent a copy of the *Crimson* to the White House every morning, but we had no idea whether the president ever read it. That summer I got a job in Washington at a small publishing company—the boss owed Dad a favor—and a bunch of undergrads who were also working in the capital decided to form a lunch group and invite our former professors. Bundy was the first guest and I got there early to grab a good seat. When I introduced myself, he said, "Roberts, you're the one who wrote that article in the *Crimson*. Maybe you'd like to know what the president had to say about it." Yes, I stammered, yes, I'd definitely be interested. As Bundy told it, Kennedy called him into the Oval Office and pointed to our story. "Mac," said the Leader of the Free World, "how could you be so stupid? You knew that if you said something like that to those guys, they'd only turn it against you."

Barely a year before, I had been writing the TV highlights and movie timetable for the *Bayonne Times*, so this was pretty exhilarating stuff, and I wanted my parents to know everything that was happening. I wrote long letters home detailing my adventures, and rereading them now, I'm struck by my parents' responses. Like me, Dad worked for a college publication, *The Square*, not the *Crimson*, and as in my case, his college years spanned a turbulent time, the late thirties instead of the early sixties. But the parallels ended there. He always had to live at home and worry about money. I could move away, confident that the tuition bill would be paid, and while Dad took great satisfaction in my accomplishments, an occasional tinge of envy crept into his letters. "How I wish I were in your milieu," read one typical note, "wrestling with the intellectual problems, soaking up the knowledge and wisdom and beauty that is your daily chore. How glad we are that you are able to do this, and that we can help make it possible."

Mom was always more practical and less sentimental. In response to one of my complaints about lack of time, she wrote: "An awful lot of good things have happened to you—perhaps too fast—very few people live in such a constant state of excitement. Learn to enjoy the more leisurely pace—it's a chance to know yourself better."

The pace didn't let up. During the spring of freshman year, I was asked to take over as the *New York Times* "stringer," or part-time reporter, on campus. It was a big break, but I almost booted it. My first assignment was to cover a crew race on the Charles River, a sport I knew absolutely nothing about. So I asked if I could ride on the judges' boat that followed the shells. A great idea, except for one thing: When the race was over, the judges didn't dock, they broke out the booze and had a cocktail party. I was frantic. I had to file my story. I contemplated leaping overboard and swimming for shore, but I was too scared. When we finally arrived back at the boathouse, a messenger from Western Union was jumping up and down and screaming: Who's Roberts? Where's your copy? On my very first story for the *Times* I'd missed the deadline. Fortunately, they didn't fire me, and I held the job for the next three years. Most of my assignments were for the sports section, and when Cokie and I started dating, we spent a lot of Saturday afternoons at track meets and lacrosse games. But I got $5 a story, and since the Saturday special at Cronin's, a college hangout near the Square, cost $1.99, the payment covered dinner for two, plus tip.

After finishing my summer job in Washington, I attended

the National Student Association meeting that produced my first article for *The Nation*. Youthful political activism was exploding all over the country, and with so many colleges in the Boston area, we were right in the center of the story. I covered a "ban the bomb" march on Washington that led to another article for *The Nation*, but I didn't always handle my success gracefully. I had two roommates who aspired to be fiction writers, and they looked down on my efforts as mere journalism. So I announced that we should devote one wall of our living room to the covers of magazines featuring our work, and we could start with my articles in *The Nation*. I'm not sure they ever forgave me. During that year I joined another student—Hendrik Hertzberg, now a staff writer for *The New Yorker*—in suggesting to the *Crimson* that we start covering the civil rights movement. The editor's reaction was "What's that?" He soon learned, as did the rest of America, and I spent most of the next decade reporting on the political upheavals ignited by my own generation. When I hit thirty, I asked my editors at the *Times* if I could finally write about grown-ups occasionally.

My social life did not keep pace with my writing life. I had few dates during my first two years at school, but one stands out. A classmate of mine, Don Bloch, was seeing a girl at Smith he knew from high school, and they had a friend who would be perfect for me. Or so they said. So one weekend, Don and I took a bus to the Smith campus in western Massachusetts. We were supposed to meet the girls at an Italian restaurant, and we got there first. There were candles in the

Chianti bottles, so I felt at home. It was a frosty night, and when our dates arrived, they had bright scarves around their necks and high color in their cheeks. I sat there thinking: This is it! Now I'm really in college! After reading Dad's college writings, the scene has even more meaning: This was the moment he'd always hoped for and never had. But the euphoria didn't last. I actually liked the girl, and we stayed in touch, but Smith was far away, and when she decided to spend her junior year in Europe, the relationship lapsed. Besides, my world was about to shift rather sharply on its axis.

During my sophomore year, I had a politics professor, Paul Sigmund, who later married Cokie's older sister, Barbara. He recruited me to attend the World Youth Festival in Helsinki, Finland, in the summer of 1962 and help publish a daily newspaper for the delegates. The festivals were sponsored by Moscow and designed to lure promising third world students into the Soviet orbit. I didn't know it at the time, but the CIA was financing a major campaign at the festival to counteract Soviet propaganda, and our paper was part of their effort. I also didn't know that the woman who lived in Paul's basement and helped organize the project, a Smith grad named Gloria Steinem, would make her own mark in the world. My parents had never traveled abroad and neither had I. In fact, I'd taken my first plane ride only a few months before, covering the Harvard basketball team on a trip to Philadelphia. To save money on the long journey to Finland, we'd been booked on Icelandic Airlines, and when my parents drove me to the airport, we couldn't find the gate. People kept directing us to increasingly remote locations, and finally we found the right spot, marked by a handwritten sign on a small blackboard. Mom and Dad almost tried to stop me from getting on the plane. But because they didn't, my life changed completely.

Another journalist at the festival, a Yale student named Bob Kaiser, kept telling me about a great girl he knew at Wellesley, Cokie Boggs. But Bob, who later became managing editor of the *Washington Post*, made a big mistake. He stayed in Europe, while I came home after the festival to attend the student association's summer meeting. And that's where I met Cokie. I was a Northern Jew whose grandparents had been born in Russia. She was a Southern Catholic whose first ancestors had come to America in 1621. Other than that, we made a perfect couple. In fact, I often joke, we both were children of immigrants—our families just arrived about three hundred years apart. A more relevant statistic: Our dorms were only 12.5 miles apart. When we got back to school, we went out a few times and were clearly attracted to each other. But like a typical guy, I got frightened. I stopped calling. Fortunately, I had several friends dating Wellesley women at the time, and they'd occasionally run into Cokie, so we never completely lost touch. The next spring, we both decided to attend a political conference in Washington and wound up traveling in the same car, driven by another student. They picked me up in Cambridge, and as I approached the car, I could see Cokie through the window, sitting in the backseat.

I immediately realized what a jerk I'd been. We've been together since that weekend.

Several of us from Harvard were supposed to stay at the Boggs house, but fortunately the others all decided not to come. Kaiser was in town, however, and after a party that Saturday night, he was angling to take Cokie home. But since I was staying with her, I had the inside track, and once we got back to her house, we spent hours talking, deep into the night. At one point Cokie made us an early breakfast, and the room where we sat, chewing over our scrambled eggs and scrambled future, is directly below where I am sitting right now, writing these words. We've lived in this house for more than twenty-seven years, two years longer than her parents did. Our daughter grew up in her mother's old room. Cokie and I still sleep in her parents' old bed. I still farm her father's vegetable patch, using some of his old tools. At Christmas we harvest Jerusalem artichokes my father-in-law brought from Louisiana and planted with his own hands. Out by the road is a flourishing clump of daylilies my father took from his own garden and brought to us. Nearby is a purple-leaved plum tree Dad helped me plant one day, and next to the lilies are three crepe myrtle trees. They originally came in pots, as decorations for our daughter's wedding in the backyard, but now they've been permanently replanted. I don't just live in my fathers' houses, I tend my fathers' gardens.

Once Cokie and I left her house and returned to school, staying together was not easy. I was a good son. I wanted my parents' approval and did everything to earn it. I followed their traditions, gratified their hopes, and came home regularly, without my dirty laundry. But they hadn't bargained for this one. Their Jewish son dating a Catholic girl was just too much. One of my visits home triggered a harshly worded, five-page, single-spaced letter from Dad. He talked about how pleased he'd been to meet the friends I'd brought with me, and then said: "Steve, visualize a family in which the father is of one faith and the mother and children are of another. Can there be the same kind of relaxed, secure atmosphere? Will a Catholic child be as comfortable bringing home his Catholic friends to meet his Jewish father?" He knew how much I respected him and he played on that: "You could not teach your Catholic children in the same way that your father taught you." Then he reminded me: "You are the boy who came to us at the age of ten and asked to be Bar Mizvahed. You precipitated the decision of this family to join a congregation, and I am glad that you did. For we all rediscovered our identity as Jews and developed a security within our Jewishness." In the end he urged me to break off the relationship: "For I can see no viable solution, nor have you suggested any to me."

I wrote back, and for the only time in my entire life, I took direct issue with my father. I resented his assumption that he was clearly right and I was clearly wrong. "Throughout the letter," I wrote, "you showed a marked insensitivity to one great factor in this matter: my love for Cokie. . . . I continue to hope because I love; you see no hope, but you do not love." I concluded: "You will say that I am young, and that my head is clouded by emotion. Of course, but that is just the point. If a relationship is to work, the energy to build a life together must come from the two people involved. And it is my youth and my love that gives me that energy."

I refused to give in to his pressure, but today I have a better understanding of his motives. He was filled with fears for me, and perhaps for himself. If my children would be uncomfortable introducing their friends to their Jewish father, what about their Jewish grandfather? When he said I could never be a "dominant male" if I married a Catholic woman, he was reflecting his own insecurities—about being a man, about being Jewish, even about being American. I didn't share those insecurities. I was another generation removed from the Old Country. I was an American, and a Jew, and I was proud of both. He was the one who had changed our name to disguise our Jewishness. I was the one who had led the family back to its Jewish origins. And it was precisely because I felt strong and secure in my identity that I could listen to my heart, and not to my parents.

In opposing my relationship with Cokie, Dad had forgotten one of his most important lessons: Think for yourself. He hated it when I wanted to act and dress like everybody else. The one excuse that would never fly with him was "All the other kids are going," or "All the other kids have one." I had a huge fight once with my mother because she bought me a V-neck sweater and all the other kids had crew necks. Or the other way around. But when I finally did listen to Dad, when I finally started thinking for myself, when I fell in love with a girl he didn't approve of, the tension was terrible. Standing up to my parents, and sticking by Cokie, was the hardest thing I had ever done. I felt a deep allegiance to my family. I was determined to carry on the work of my father and grandfather and uncle. But I had to do that with the person who was right for me, who would give me the strength and support to shoulder that burden.

Years before, in describing their first date, Dad told Mom that he knew she would "make it easy" for him. I understood exactly what he meant. That's how Cokie made me feel. She made everything easy, and exciting. If the arrangements weren't perfect, if the movie was dull or the coffee cold or the train late, it didn't matter. Finally, I could relax and be myself. And to his credit, Dad eventually realized his mistake. After he got to know Cokie better, he conceded that it was hard to keep opposing our match because she was obviously "the perfect girl" for me. Once I broke down the taboos I was raised with, I was ready to take up the obligations that came from that same upbringing. But I needed the right partner at my side, even if she was from another tribe.

Three people made my reconciliation with my parents easier. One was my brother Marc, who went to them and said: Cokie and Steve are going to get married, you're not going to stop them, all they want is your blessing. If there's any rift here it will be your fault. Mom and Dad thought of themselves, with good reason, as such loving parents that Marc's words shocked them deeply. My future mother-in-law, Lindy Boggs, immediately made my parents feel welcome and eased a lot of their concerns. And when we got married, two years after Dad wrote that letter, Lindy gave up her dreams of a church wedding and agreed to let us get married at home, a setting where my family and their friends would feel more comfortable. But it was Cokie who really clinched it. She became, as my parents eventually acknowledged, the "best Jew in the family." As a believing Catholic, she took religion and ritual seriously, and made them central to our life together. Early in our marriage, we started having an annual seder, the festive ritual that Jews celebrate at Passover, and when we eventually returned to Washington, and moved into the house where we had been married, my parents became regular and enthusiastic guests at these events. I knew that any residue of resentment had completely disappeared when Dad said, Don't even bother asking us anymore, just assume we'll always be there. As I was writing this chapter, Cokie and I attended Yom Kippur services together. And as I heard her singing the Hebrew prayers, in her strong, clear voice, I thought of Dad and wished he could have heard her, too.

I thought of Abe as well. He, too, came to love Cokie, and for a simple reason. She was kind to him. She paid attention. I have several letters he wrote to her, and in one from 1967, a year after we were married, he says how pleased he is that we're planning to visit him in Bayonne. He tells her that he will serve "the choicest club steaks that a prize steer ever produced" and a litany of succulent fruits, from "big black grapes from Cal" to "D'Anjou pears from Oregon." He will even have "giant persimmons," the fruit, he assures her, that

Adam really shared with Eve. "Please let me know as to the date and hour," he asks, "so I can have the steaks seasoned and the fruit ripened." Like Dad, he realized that I'd found "the perfect girl."

A year later, when we had our first child and named him Lee, after Uncle Bussy, Abe was deeply touched, and we invited him to a party at our apartment in Manhattan to celebrate. Age had not erased his eccentricity. The morning of the party he called and said he couldn't come. Why, I asked, keenly disappointed. There won't be a place to park, he explained. Listen, Pop, I assured him, call when you're leaving. I'll lie down in the street and save you a spot. He finally did come, and a few days later wrote to Cokie again: "The more I see of you, the more I admire and respect you. I hope my other grandchildren will be as lucky as Steve in finding their proper mates." A few months later, we left for California and never lived near Abe again. But we kept him supplied with pictures of the kids, and when Lee was almost a year old we got a note from him, with these words: "When I come home after many heartaches, headaches, disappointment and problems galore, I drop in my chair and start reading the New York Times, then somehow my mind wanders off in mulling over my predicaments. But if my gaze happens to fall on Lee's picture, I start smiling, and I forget about everything." When Abe was dying in 1977, and he didn't always recognize members of his own family, he would ask for Cokie. She'd made it easy for him, too.

EPILOGUE

CROSSING THE RIVER

In the fall of my senior year, a brief conversation at the *Crimson* changed my life in yet another way. Don Graham, whose family owned the *Washington Post*, told me he'd had a summer job working for James Reston, the Washington bureau chief of the *New York Times*. You should apply for it, he said. When I wrote to Reston he replied: I took Don for the summer as a favor to his family, but I'm really looking for someone to work a full year as my research assistant. Would you be interested when you graduate? Yes, I would, and on November 1, 1963, I flew to Washington to meet with him. For luck I wore my *Crimson* tie, black with red stripes, and after our chat, Scotty introduced me around the newsroom. Born in Scotland, he'd grown up in Ohio, attended the University of Illinois, and harbored a healthy skepticism of Ivy League snobs. When we approached Anthony Lewis, the

Times's Supreme Court reporter, Scotty mentioned that I was attending Harvard, Tony's old school. "I know," Lewis replied, "I recognize the tie." Not a good moment. But by this time I had allies: the sports editor of the *Times*, who had forgiven my first botched assignment; Cokie's mom, who was a friend of Scotty's; and my housemaster at Harvard, John Finley, whose father had once run the *Times*'s editorial board. Everybody, Scotty once cracked, recommended this kid except Charles de Gaulle. Three weeks later President Kennedy was killed. Cokie and I wanted to share our grief together, and retreated to a friend's cabin in New Hampshire. Kennedy had been elected in the fall of our freshman year, and his

My former boss, the New York Times *Washington bureau chief, James "Scotty" Reston, with several of his former clerks. Scotty is in the center, with the bow-tie. I am behind him to his left. This was taken in 1984, twenty years after I began working for him.*

death, in a way, ended our senior year months early. We were Kennedy's children, and it was time to grow up. A month later, when I was home for the holidays, I got a handwritten note from Scotty saying, "Here's another Christmas present, you have the job."

My freshman year had been an academic disaster. I hated French and a required course in the history of science. (A fellow freshman who took that course with me became a writer of scientific and medical novels. I guess Michael Crichton got more out of the subject matter than I did.) But by the middle of sophomore year, something clicked, and my grades improved. I studied with a lot of famous professors, but the man who made the biggest impact on me was a young government instructor named John Rodman. Under the Harvard system, you were assigned a tutor in your academic major and during junior year you met regularly, one on one. Rodman pushed me hard, saying repeatedly after reading my work, "That's not good enough." It grated at the time, but I'm deeply grateful now and I find myself repeating Rodman's words to my own students. As senior year approached I had a decision to make: Compete for a top editing job on the *Crimson* or write an honors thesis. I didn't feel I could do both well, and since I knew I would spend my life working for newspapers, I chose the thesis, but I didn't stray far from my career path. I wrote about the political philosophy of Walter Lippmann, the first modern newspaper columnist. As I was reading about Lippmann's early years, I was stunned to discover that after graduating from Harvard in 1910, he'd moved to Washington as a research assistant for Lincoln Steffens, the great muckraking journalist. And here I was, fifty-four years later, preparing to become an assistant to Scotty Reston. My fantasy of following in Lippmann's footsteps kept me going through the long Cambridge winter, buried deep in the musty stacks of the Harvard library system. Writing a thesis was a brutal experience, but today I urge my best students to try it. Nothing else teaches you so much about your own talent or tenacity.

I moved to Washington a week after graduation and spent the next twenty-five years on the *Times*. Cokie moved there as well, living at home and working at a small TV production company. We had spent many tearful nights during our senior year, wondering and worrying whether we could ever make our relationship work, but we were determined to try. Mom sent me off with a supply of cold cereal, apparently unaware that glazed donuts were readily available in the nation's capital, and I never finished the cornflakes or the raisin bran. Dad, as usual, was less practical and more excited. He was eager to share and shape the life I was living, and when he discovered that Reston was a fellow pipe smoker, he sent him the batch of Walnut tobacco I mentioned earlier and a note about their membership in the "pipe-fidgeting fraternity." In a letter to me, Dad revealed his feelings: "I must confess that I spent a great deal of time composing that nonchalant little note. I felt that perhaps I should include some profound thought or meaningful insight which would hopefully be the springboard of a column, but I was unable to summon up any." Dad also liked a column by Russell Baker, who occupied the office right next to mine, about the "slob syndrome." He wrote to Baker that his words had "struck a responsive chord in millions of gravy-stained breasts." Dad described himself as a proud member of the "Great Rumpled Mass" and warned Baker: "Don't judge me by that neat young man who works on your floor and bears my name. He looks tidy when he hasn't been near a barber for months. He takes after his mother. I look disheveled ten minutes after I left the barber shop." The truth is, I took after them both. Dad kept me going and Mom kept me grounded. Her notes from that year include this admonition: "I hope you're being judicious about putting some of your salary aside." The advice was unnecessary. I'd started putting money aside when I was making quarter tips delivering the *Bayonne Times*. I was her child. How could I do anything else?

That year in Washington drew me deeper into a new world. Since Cokie was living at home, I visited there often and got to know her parents well. They often included us in dinner parties with their notable friends, and one night Scotty called and interrupted the gathering. The *Times* had secured a document that could have great meaning for America's relationship with the Soviet Union, and he wanted me to fly to New York immediately and pick it up. One problem: I had no credit cards and no cash for airfare. I approached my future father-in-law and several of the guests, including at least one cabinet officer. None of them had any money, either. Finally, I scraped the fare together and raced to the airport, but missed the last plane. I took the first flight the next morning, picked up the document, and delivered it to the White House. It turned out to have minor significance, but that wasn't the point. The incident offered a brief glimpse into my future, and two decades later I covered the White House for the *Times*.

Cokie's parents were good friends with President and Mrs. Johnson, and Lady Bird would invite us occasionally to White House functions. She called one day in midafternoon. There was an outdoor reception that night following a state dinner for a visiting African potentate, a few places had

The Associated Press took this picture of us on our wedding day, September 10, 1966. From left: *Cokie's sister, Barbara Sigmund, Cokie's Dad, Hale Boggs, and President Lyndon Johnson and Mrs. Johnson. We got married in the garden of Cokie's home to make my family feel more comfortable, and we've now lived in that house for the last twenty-seven years.*

turned up at the last minute, and would Cokie and her beau like to come? I didn't own a tuxedo and it was too late to rent one, so I scrambled around, assembling pieces of an outfit from various friends. One problem: The studs didn't fit the tux shirt. And as Cokie and I stood in the receiving line that night, with the president and his guest advancing toward us, the studs kept popping out. Then the heel of Cokie's shoe broke. So the first time I ever met a president, I was holding my shirt closed with one hand, while my date was hopping on one foot. Clearly, I wasn't ready for prime time.

But during that year, I did learn to love and admire two surrogate fathers. They never replaced my own dad, but they expanded and enriched my experience. The first was Cokie's father, Hale Boggs. He was then the majority whip of the House, the number three Democrat, but he loved meeting his children's friends and always took me seriously. He was a Southerner, but he believed deeply in racial equality, and he risked his political career to support a bill that guaranteed voting rights for blacks. If my children and grandchildren remember anything about Hale, who died in a plane crash in 1972, I want it to be that one act of supreme honor and courage.

Scotty Reston became a model and mentor in many ways, and everyone who ever worked for him had the same experience. When you were new on the job, he'd charge into your office, pipe ashes flying, and ask you to read the column he'd just written. The usual reaction was, "It's brilliant, sir." So he'd laugh and say, "Look, I can get anybody to answer my phones. You're only a help to me if you tell me what you really think." So maybe the next time you'd timidly suggest changing a word or two, and he'd do it! It was a great lesson. He'd take advice from a clerk as readily as from a congressman. He was also a journalist of virtue and grace, incapable of betraying a confidential source or writing an awkward sentence. And he was Cokie's biggest fan, asking me often during our year together, with increasing urgency: "When are you going to marry that girl?" The fact that he could have a luminous career, a loving marriage, and three children made a profound impact on me. I'd always wanted to be a father. I had a good one and I wanted to be like him. I also wanted to be something that Dad never was, a writer. Scotty showed me that I could be both. After the tie incident with Tony Lewis, however, it took me a while to erase his image of me as an arrogant Ivy Leaguer looking down on his Big Ten pedigree. One evening I arranged to deliver a batch of mail to his home, and when he invited me to stick around for a drink, I got the chance to tell him about my background, about Bayonne and Grandpa Abe and my own immigrant roots. I'm like you, I told him, not like them. After that conversation his attitude softened and our relationship relaxed.

My primary job was answering Scotty's mail and doing research for his columns, but I quickly realized that the star reporters assigned to major beats left a lot of other subjects uncovered. So I started contributing articles to Sunday feature sections like travel and entertainment, and going to the office every Saturday, just in case there was an extra story to do. On one of those Saturdays, right on deadline, someone threw something at the Soviet embassy, a can of paint, as I recall. Since no one else was around, the desk editor sent me to check it out, and that incident produced my first front page story. After a while, the weekend editors started giving me regular assignments. In addition, the civil rights story was erupting. During the summer of 1964, rights workers flooded Mississippi and three were killed. The Washington bureau had no one assigned to that beat, and since I had been covering civil rights for the *Crimson*, I had a lot of good contacts and wound up doing a number of stories for the *Times*. By the end of the year, I had a fat packet of clippings and Scotty's support for a reporting job on the city staff in New York. But New York resisted. I was twenty-two, far younger than any of their other reporters, and as one editor told Scotty, he had the most promising journalists in the country knocking down his door. Finally, Reston got exasperated. He flew me to New York, marched me into the office of managing editor Clifton Daniel, and did some knocking of his own. "Are you going to hire this kid or not?" he demanded. A few hours later, Daniel reluctantly relented. I had the job. That's why Scotty's picture is on my office wall. And that's why, whenever I meet a new class, I tell his story. When I worked for him, I remind my students, he was the most prominent figure in American journalism. And yet every day, he took time for me—to answer my questions, read my stories, boost my confidence. So if you're grateful for something that happens here this semester, I ask only one thing in return. You help somebody else.

It works. Young people usually live up to the expectations adults have of them. And it pleases me when some of my students refer to me as "their father away from home." As my own kids always say, two children weren't enough for Dad, so now he has dozens of them. In fact, my own father had a similar impulse, four kids weren't enough for him, and after we left home, he used to recruit young people to serve as his crew in sailing races. Dad, like Abe, had his quirks, and one of them was his sailing attire. Shorts left his legs cold but long pants got wet, so he had Mom cut down some old trousers at midcalf, sort of homemade clamdiggers or pedal pushers. As a tribute to Dad, his crew once showed up for a big race decked out in similar gear. I love the picture of him and his young friends, all sporting their peculiar cutoffs. He died suddenly, in the middle of the night, so word hadn't spread. And the day after his death, the crew showed up at the house, ready to take him out for a race. He was a dad till the end.

I was in Cleveland not long ago, giving a speech, and the mother of a former student approached me. She wanted to show me the picture of her daughter that she keeps on her desk. There was Gayle, and next to her there was me. People who come into my office, the mom said, laughing, think you're Gayle's father. I know Gayle's real dad, he's a great guy, and I'm happy to serve as his occasional substitute. That's what Hale and Scotty did for me. After Scotty died in 1995, he taught me a final lesson. He had twenty-six clerks in all, and eighteen of us served as ushers or pallbearers at his funeral. He understood what his real legacy was. And as I helped carry his coffin from the church, I understood it, too. As I told his widow, Sally, his life had been like a pebble tossed into a pond. The ripples of his influence would keep extending outward, touching future generations that would never even know his name.

A Roberts family portrait taken at my parents' fiftieth wedding anniversary in 1990. Dad used to look around at such gatherings and say, "See what happens when you walk a girl home from a party."

After leaving Washington, I stayed with my parents in New Jersey for a few days, and the evening before I started work on the city staff, I took the bus into New York. As we approached the entrance to the Lincoln Tunnel, I could see the whole West Side of Manhattan spread out before me. "The city" that had loomed so large in my childhood, just across the Hudson River and yet so far away, would be my home, and my subject. I was a real writer now. But crossing over the river did not mean crossing out the past. Dad and Mom, Abe and Bussy and Scotty, were all on the bus with me.

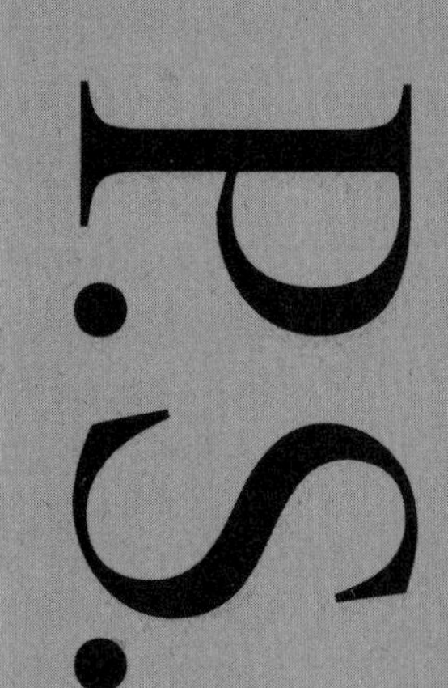

Insights, Interviews & More . . .

*

About the author

About the book

Read on

About the author

Meet Steven V. Roberts

Jan Cobb

"Roberts covered some of the major events of his time, from the antiwar movement and student revolts of the sixties and seventies to President Reagan's historic trip to Moscow in 1988."

STEVEN V. ROBERTS has been a journalist for forty years. Roberts covered some of the major events of his time, from the antiwar movement and student revolts of the sixties and seventies to President Reagan's historic trip to Moscow in 1988 and nine presidential election campaigns. After graduating from Harvard *magna cum laude* in 1964, he joined the *New York Times* as research assistant to James "Scotty" Reston, then the paper's Washington bureau chief. His twenty-five-year career with the *Times* included assignments as bureau chief in Los Angeles and Athens and as congressional and White House correspondent. He was a senior writer at *U.S. News & World Report* for seven years, specializing in national politics and foreign policy. Roberts and his wife, television journalist Cokie Roberts, write a nationally syndicated newspaper column and are contributing writers for *USA Weekend,* a Sunday magazine that appears in five hundred newspapers nationwide.

In February 2000, William Morrow published *From This Day Forward,* an account of Steve and Cokie's marriage, as well as other marriages in American history. The

New York Times called the book "inspiring and instructive," and it spent seven weeks on the *Times* bestseller list.

A well-known commentator on many Washington-based TV shows, Roberts also appears regularly on the ABC radio network, serves as cohost of the public radio show *America Abroad,* and is a substitute host on NPR's *The Diane Rehm Show.* As a teacher, he lectures widely on American politics and the role of the news media. Since 1997 he has been the Shapiro Professor of Media and Public Affairs at George Washington University, where he has taught for the last fifteen years.

His many honors include the Dirksen Award for covering Congress, the Wilbur Award for reporting on religion and politics, the Bender Prize as one of George Washington University's top undergraduate teachers, and four honorary doctorates. He's been named a Father of the Year by the Father's Day Council and received the Public Service Sector Award from the Aspen Institute. Steve and Cokie have two children—Lee, an investment banker in New York, and Rebecca, a journalist in San Francisco—and six grandchildren.

In his spare time Roberts is an avid gardener and tennis player.

"Roberts has been named a Father of the Year by the Father's Day Council."

The Afterglow of Publication
Hearing from Readers

"HEY, I DID THIS, TOO!"

"I also dream about Bayonne, and my 'Block' was also my world." Those are the words of Mara Lynn Chertoff Steinberg, who grew up around the corner from me on Thirty-second Street in Bayonne. I have been deeply touched by the reactions to this book, particularly from readers like Mara Lynn who feel their own lives and families are reflected in these pages. I tell my writing students to value the "nod factor," and I don't mean putting people to sleep. Aim for a nerve of memory or emotion, and cause your readers to nod in recognition. So I was thrilled when my friend Amy Isaacs wrote: "Abe could have been my Uncle Howard the roofer. Bayonne could have been Lansdowne, Maryland. My grandparents on my mother's side came from Russia, too. As I read your book I could hear the accents. I laughed. I cried. I drove others nuts reading passages to them."

"I have been deeply touched by the reactions to this book, particularly from readers who feel their own lives and families are reflected in these pages."

Many friends and neighbors from Bayonne connected with these stories and added details of their own. I was speaking in a New York bookstore one evening and I told the tale of returning to the Block and encountering a woman who demanded to know if I remembered her. In trying to figure out who she was I started counting up from our house at number 174. The Lautons were at 172, the Simons at 170, etc. In the audience a hand shot into the air. "I'm Joyce Lauton from 172," announced the woman. In Bayonne addresses were so important because they marked your turf and your tribe. Elliot Moritz, who lived on the other side of us at 176, wrote: "On page 109 there's a picture of a Halloween party in your backyard (where we shot hoops). I think that's me to the left of the baby on the swing. I'm almost sure of it." He's right. He's the one in the clown suit and the mustache. Joyce is in the picture too, wearing a headband and baggy pants. She's supposed to be a gypsy. Judi Goodman Grant sent her own photo—she and her twin brother Bobby in matching outfits at age three—and described her own roots at 169: "My parents, Melvin and Ross Goodman, and my grandma Lilly Lang were living there when I was born. My mother told me the story of how when my twin brother and I were born your mom sent over flowers from you and Marc welcoming the new set of twins on the Block." In the photo Judi and Bobby are definitely standing on West Thirty-first Street. I don't recognize them, but I do recognize the front stoops in the background.

One of the most intriguing messages came from Kathie O'Donnell, who used this subject line in her e-mail: "174 West 31st Street—My home." She and her husband John bought our old house ten years ago. "As far as my family is concerned, we have six children and are very happy here," she wrote. "Both my husband and I are Bayonne natives and decided to raise our children here." I've happily accepted her invitation to come visit. But six kids? In that house? Even though her oldest has left home to join the Marines—where he probably has more privacy than he did at 174—I know how small that place is. We moved out after my sister Laura—child number four—was born. She had to sleep on the porch.

Grandpa Abe was a favorite topic. ▸

"One of the most intriguing messages from a reader came from Kathie O'Donnell, who used this subject line in her e-mail: '174 West 31st Street—My home.'"

The Afterglow of Publication ***(continued)***

Samantha Drogin recalled that her family home on Roosevelt Terrace was "tushy to tushy" with the house Abe built on Lincoln Parkway. She failed to mention whether her father, like my grandfather, had illicitly filled in Newark Bay to widen his building lot. Mara Lynn Steinberg also had memories of that house: "My father's sister Esther Chertoff Rosenthal and my Uncle Leon were tenants of your Grandfather Abe on Lincoln Parkway. My aunt often spoke of how your grandfather made passes at her. The family thought she was exaggerating, but after reading your book I believe it could have been true. I remember the house very well. It was more upscale than most two family homes in Bayonne. I played there often and never realized that the ground surrounding the property could have fallen into Newark Bay!"

Sunset Trailer Park could have fallen into the bay as well. But like the house on Lincoln Parkway it survives today, defying all laws of geology. Since Abe created most of the land out of sunken barges and truckloads of rubble the park probably should have washed away years ago. As a young bride in the mid-fifties Katherine Kemner lived there while her husband attended college in Jersey City. After his death Katherine married a man she had known back at Sunset and she reminisced: "We have often talked of those days in Bayonne and many times have wondered if Sunset Trailer Park still existed or if it had sunk into Newark Bay. It was almost a betting proposition among some of the residents there as to what part would sink first, a special concern of mine since the first year there we had a small old trailer at the very end, right on

"We have often wondered if Sunset Trailer Park still existed or if it had sunk into Newark Bay. It was almost a betting proposition among some of the residents there as to what part would sink first."

the bay! We were rather surprised to know that that part of our past still stands."

Abe's antics reminded a number of readers of their own relatives. Marjorie Harelick, a high school classmate, recalls her postal clerk father: "He was a character like your grandfather Abe. Job Lot was his favorite store and he brought home merchandise in bulk like mink skulls, which he made into pen and pencil sets and tried, unsuccessfully, to sell." I'm stunned. The mink skulls didn't sell? Perhaps, like the shower curtains Abe once bought by the gross, remnants of Mr. Harelick's dream are still moldering somewhere in Bayonne.

Some of the most heartwarming letters came from readers who remembered my Dad's books. Jim Johnston wrote from Richmond: "My clearest memory as a child was being read to by my parents. You can imagine my excitement when I learned on page 125 of your book that one of my favorite books, *The Fix-It Book,* was designed and written by your father. I loved having that book read to me. In fact, I dragged that book around so much the handle finally deteriorated and tore off. But mom and dad still kept reading it to me." Jim clearly recalls the brown cover and the vivid accounts of electricians, mechanics, carpenters, and truck drivers. He even thinks Dad's book influenced his career choice, since he spent thirty-two years recruiting and hiring similar workmen for a large company. I still have a copy of *The Fix-It Book.* The binding is ripped but the handle is intact.

Friends of my siblings checked in as well. One woman admits to having a "crush" on ▸

my brother Glenn. Another recalls exchanging "steamy" letters with my brother Marc in high school German class: "I kept them in my night table tied with green ribbon!" Funny, nobody wrote confessing a crush on me. Mark Berman did recall that a friend of his once dated Renee Cherow, the girl I longed for all through high school. "You may have been shy," he offered, "but your taste in women was [and is] impeccable." I'm sure Renee, and my wife, appreciate those sentiments.

> "Friends of my siblings checked in as well. One woman admits to having a 'crush' on my brother Glenn."

Many readers who related to the book also belong to Jewish immigrant families. June Wallach had an aunt who married a man named Louis Rochowitz. He changed his surname to Roberts and had a son named Steven. Actually I know several Steve Robertses whose original names resemble Rochowitz or Rogowsky. Marty Donsky, another Bayonne boy, wrote: "On almost every page I found myself saying to my wife, 'Hey, I did this, too.' I know it is your family's story. But it is very much my family's and every other Jewish family's story." Letters like that one reinforce the adage I tell my writing students: there are two kinds of great stories. Some hardly ever happen. Others, like the ones in this book, happen every day. I want people to say, "Hey, I did this, too."

While touring for this book I spoke at the Jewish Community Center in Atlanta. They were mounting a production of *Fiddler on the Roof,* so I was actually standing on the set of the Russian village, Anatevka, as I made my remarks. That night reminded me that when *Fiddler* was produced in Japan it was a huge hit. The audiences didn't know a thing about Jews in Russia a hundred years ago, but they knew a lot about fathers and daughters and families disrupted by changing times. The themes of *Fiddler* were universal. They told us something about human nature. So I was particularly pleased when readers who don't share my origins responded to these stories.

One of them is a Catholic monk, Brother Mark De Brizzi, who grew up in Bayonne about fifteen years after me and lovingly captured the town's culture: "When you are born in Bayonne you are born into a tribal dance. And each tribe teaches you a few dances. What do I know from lox and bagels? Mr. Chas. Stern taught me from such things! The Polish taught me how to buy good sausage and pierogi and the Italians taught me how to wail loudly at funerals. The older Jewish neighbors taught me [how] to think of FDR (who 'saved us from the Depression') and how to tip my hat to a lady when I got on a bus." Indeed, adds the good monk, "Bayonne does tend to stick with you, the same way the floor of the Dewitt Theater stuck to one's shoe."

"Bayonne does tend to stick with you, the same way the floor of the Dewitt Theater stuck to one's shoe."

As Br. De Brizzi notes, food often defines a community in a town like Bayonne and that theme runs through many messages. One recent visitor reports that the pizza served at the Venice is "the most delicious I've had in forty years." Doris Slavin recalls that "the most delectable sundae savored in my life was the gooey marshmallow chocolate syrup served in Bayonne." Several readers insist that the orange sherbet and vanilla ice cream treat I describe in the book is a Creamsicle, not a Dreamsicle. The two confections, both sold by Good Humor trucks, are often confused. But I've researched this carefully and I'm ▶

“Churches, too, were central to Bayonne life, and you practically needed a visa to cross the frontier from one parish to another.”

convinced that what I remember is a Dreamsicle. If you want to see a picture of a Dreamsicle or to buy a refrigerator magnet featuring that orange and white delight go to www.famousfoto.com/tin-signs/popsicle-magnets.htm.

Because Bayonne was such an isolated and self-contained place, people were deeply devoted to their local hangouts. One of them was the DeWitt Delight Luncheonette, run by a Greek family that changed its name from Thomopoulos to Thomas. Peter Thomas, the son of the owner, recalled that the weekday lunch crowd was dominated by local business people: “In your book you mentioned Penner's store [the Penners lived across the street] and I began to laugh. Penner was a human jumping jack, who entered our store armed with jokes and one-liners.” Another regular was Neil Gallagher, a local lawyer who got elected to Congress and ended up in Federal prison. Thomas described him: “With that handsome face, Irish smile and blue eyes, and always impeccably dressed, he had everything going for him.” But Gallagher fell in with the mobsters who were always part of Bayonne's seamier side. “The town was crawling with gamblers,” says Thomas, “and we knew damn well that some of our evening customers could well have been extras in the *Godfather* opus.”

Churches, too, were central to Bayonne life, and you practically needed a visa to cross the frontier from one parish to another. But Father Benedict Groeschel said there were a lot of relationships in town that spanned religious and ethnic boundaries. Because he lived near many Jewish families growing up, wrote the priest, "I've always had a strong place in my heart for the Jewish people and things Jewish." At times he even served as a "Shabbas goy," a Christian who helped with the tasks observant Jews are barred from performing on Saturday.

Hearing from so many readers who want to share their own memories reinforces an important point: every family has a story worth preserving. So I was delighted by the note from George Krill, who brought his son Alexander to hear me speak at the National Book Festival on the Mall in Washington: "He now wants me to record my family's history after seeing and listening to you." When my mother went down into her basement and found that box of letters my parents had exchanged more than sixty years before she was giving me a priceless present. Every family should have a box in the basement, figuratively if not literally. One reviewer said my book read as if I had written these stories down for my grandchildren. She clearly was not paying me a compliment, but my reaction was: so what's wrong with that? I hope you will fill your own box and give your family the gift of their own history. Here are some tips for doing that.

First of all, pursue and preserve every shred of evidence from the past. George Krill, for example, sent me copies of the Czech-language travel documents his parents used when they immigrated to America. They contain a lovely picture of his mother wearing a traditional headscarf. They also contain specific dates and places, which are extremely useful in outlining a family history. When I found a photo of Abe and Miriam dated ▶

The Afterglow of Publication *(continued)*

> “Look for relics in old trunks, drawers, or file cabinets. People forget they have them or don’t think they’re important.”

Bialystok, 1912, I could build a chronology in both directions using that moment as a keystone. Look for relics in old trunks, drawers, or file cabinets. People forget they have them or don’t think they’re important. Ask relatives and friends for any artifact that might be useful—not just letters and photos, but souvenirs, gifts, books, passports, records, or awards. Anything at all, particularly if it has a date or a place attached. I have ration books my parents used during World War II and copies of their earliest tax returns. My great-niece owns a volume inscribed by her two grandmothers, who were college chums during the Depression and cut expenses by sharing books. A student of mine found a cache of letters written by her great-uncle during World War II. They start with the bravado of youth as he heads off to flight school and end with a growing sense of doom as he prepares for his last mission over occupied France. Another student uncovered a journal written by her grandfather while he worked for *Ripley’s Believe It or Not!*, traveling the world to document the strange occurrences described in that popular newspaper feature. A third student rescued poems her grandfather had written back in Pakistan and translated them into English. Once you’ve located these treasures, keep them safe. We’ve salvaged countless photos from old albums and had them reprinted on special acid-free paper. My parents’ letters are now stored in watertight plastic boxes on the second floor, far from ground-level floods that can destroy a family legacy overnight.

Technology can be the enemy of history. Most people don’t write letters anymore. They trade e-mails or instant messages or simply talk on their cell phones. It's a good idea to download important e-mails and save hard copies or back them up electronically on discs or CD-ROMs. The same goes for digital photos. It's fun to get them on your computer, but print out the best ones and catalogue them in albums. Your grandchildren will thank you. Technology, however, can also be a great asset in searching for your family roots. Numerous databases are now available and more are coming online all the time. One example: after this book was first published my friend Joe Ryan, an ardent student of civic and family history, discovered something called "the old man's draft" registry. It was compiled in 1942, and all males between forty-five and sixty-five were required to sign up. After finding his own grandfathers, Joe looked up mine, and their forms were a revelation. I learned for the first time that my Grandpa Harry came from Grodno. The town was a provincial center of Jewish life not far from Bialystok, where my other grandfather Abe was born. Both listed "Russia" as their home country, but Grodno is now in Belarus and Bialystok in Poland. I was also surprised by their size. Harry listed himself as five feet six inches tall and Abe was only an inch and a half taller. I'm six feet and I remembered them as bigger. But the forms also raise a warning flag. It's easy to get confused by names. Many of them have been changed or misspelled on documents, so try different variations when you're scanning a database. For instance, on his draft form Harry lists his own daughter's name as Dorothy Rogoff. That's wrong. It was Rogow. I've also found census data for 1920 and ▸

"Technology can also be a great asset in searching for your family roots. My friend Joe Ryan, an ardent student of civic and family history, discovered something called 'the old man's draft' registry."

> “Even more precious than documents and records are personal memories.”

1930 online, and those listings are particularly useful because they provide a lot of facts. But again, beware. Abe Rogow was easy to locate. But Harry's last name, Schanbam, was misspelled, perhaps by the census taker. It took my buddy Joe many attempts using different spellings before he found the right listing.

One place to start your own search is Ancestry.com, which can then link you to various databases such as the census and the Social Security system. Joe Ryan suggests mining the “enormous collection of data” gathered by the Mormon Church. Ellis Island has voluminous records of refugee arrivals from all countries. For Jews, jewishgen.org is a priceless resource. Joe points out that you have to track down every lead. He remembers watching a TV documentary and noticing that one of the experts on the show had his grandmother's maiden name. He located the man through the producers and sure enough, they were related. Joe also found a relative in Ireland through the Internet who has organized DNA testing to establish a definitive family tree.

Even more precious than documents and records are personal memories. The living history of every family is disappearing a bit each day. A neighbor on Thirty-third Street, Harvey Greenberg, wrote to say that his mother had died suddenly last summer. “Alas,” he said, “it is now too late to ask so many questions.” Don't wait until it's too late. Start talking to your own relatives right now, and there are some guidelines. Make an appointment, set time aside, and remove distractions. Don't do it on the fly or after

Thanksgiving dinner. Use a tape recorder or even a camcorder. And prepare well. Read as much as you can beforehand. Ask specific questions. Use aids like letters and photos to prompt fading memories. "Where and when was this snapshot taken?" is a better question than, "So, did you ever take a summer vacation?" You might encounter resistance from some relatives who don't want to talk about the past. As Harvey noted: "My Dad, who passed in 1996, never responded to questions since like many in your book he wanted nothing to do with his immigrant roots." Don't let your relatives off the hook. Keep pressing for answers. My students find that if they tell their grandparents the questions are for a class assignment they get better cooperation. If you can't sell the interview as an educational endeavor tell your relatives that future generations deserve to know the family history. Because they do. I am often asked why I wrote this book, and my wife Cokie provided the best answer. "Steven," she told me, "you have honored your ancestors." I hope so. Everyone who reads these words also has ancestors worth honoring. So save their stories, too.

Author's Picks
Favorite Memoirs

When people remember their own lives I find that childhood usually provides the freshest material. As they describe their adult experiences writers are drawn relentlessly toward self-importance, determined to emphasize the vital roles they played, the notable celebrities they befriended, and the brilliant insights they provided. So my favorite memoirs tend to focus on the years before the authors became famous and fatuous.

***Growing Up*, by Russell Baker**

In a wry, self-deprecating style, the former *New York Times* columnist recounts his early years in Virginia and New Jersey and draws a vivid portrait of a courageous widowed mother who constantly lectured: "Make something of yourself, Russell." He uses letters to his mother from an aspiring suitor to capture the bleak emotional climate of the Depression years, and that encouraged me to employ my parents' letters from the same period in a similar way.

***A Walker in the City*, by Alfred Kazin**

This renowned literary critic is a generation older than I am and closer to the immigrant experience. His story of a Brooklyn childhood in the 1920s describes the process of becoming American with clarity and warmth. At one point his mother explodes at a group of young women who are talking about falling in love. "What is this love you make such a stew about?" she asks, summing up the tension between Old and New World views of marriage. Reading this memoir many years ago first sparked my interest in the form.

Big Russ and Me, **by Tim Russert**

How does a famous TV star write about a father who collected garbage? This book could have easily adopted a smug or patronizing tone, but Russert's genuine respect and affection for his Dad seeps into every page. I particularly enjoyed reading about an ethnic town like Buffalo—a bigger version of Bayonne—from the perspective of a Catholic contemporary.

Wait Till Next Year, **by Doris Kearns Goodwin**

As a lifelong baseball fan I loved this account of how devotion to the Brooklyn Dodgers bound father and daughter together. During day games young Doris meticulously keeps score and recounts the action, play by play, when her dad returns from work. Many father-son memoirs revolve around athletics, but this book makes the trenchant point that girls also used sports to relate to their dads.

In My Father's Court, **by Isaac Bashevis Singer**

Born in Warsaw in 1904, this Yiddish storyteller recalls his boyhood as the son of a rabbi who often adjudicated small disputes in the Jewish community. One passage I've never forgotten: as Singer's father offers his verdict all parties hold a corner of a white handkerchief to signal their agreement. That handkerchief is the perfect metaphor for the rule of law. I read this book aloud in nightly bites to my own children.

" Reading *A Walker in the City* many years ago first sparked my interest in the form. "

Have You Read? More from Steven V. Roberts

FROM THIS DAY FORWARD

After thirty years together, Cokie and Steven Roberts know something about marriage, and after thirty distinguished years in journalism they know how to write about it. In *From This Day Forward,* Cokie and Steve weave their personal stories of matrimony into a wider reflection on the state of marriage in America today.

Here they write with the same conversational style that catapulted Cokie's *We Are Our Mother's Daughters* to the top of the *New York Times* bestseller list. They ruminate on their early worries about their different faiths—she's Catholic, he's Jewish—and describe their wedding day at Cokie's childhood home. They discuss the struggle to balance careers and parenthood, and how they compromise when they disagree. They also tell the stories of other American marriages: that of John and Abigail Adams, and those of pioneers, slaves, and immigrants. They offer stories of broken marriages as well, of contemporary families living through the "divorce revolution." Taken together, these tales reveal the special nature of the wedding bond in America. Wise and funny, this book is more than an endearing chronicle of a loving marriage—it is a story of all husbands and wives and how they support and strengthen each other.

"[T]he example of the Robertses' own marriage, and the concessions and commitment they have both obviously brought to it, is ultimately instructive and inspiring." —Michele Orecklin, *New York Times Book Review*